*How to interpret
statistical data:
a guide for librarians
and information scientists*

How to interpret statistical data:
a guide for librarians and information scientists

I S SIMPSON
MA CPhys MInstP FIInfSc

Formerly Principal Lecturer
Department of Librarianship and Information Studies
Newcastle upon Tyne Polytechnic

THE LIBRARY ASSOCIATION

LONDON

Published by
Library Association Publishing Ltd
7 Ridgmount Street
London WC1E 7AE

First published 1990

British Library Cataloguing in Publication Data

Simpson, Ian S. (Ian Stuart), *1927*–
 How to interpret statistical data.
 I. Title
 519.5

ISBN 0-85365-729-7

Typeset in 10/12pt Times by Library Association Publishing Ltd
Printed and made in Great Britain by Bookcraft (Bath) Ltd

CONTENTS

LIST OF FIGURES

AUTHOR'S PREFACE

In teaching students of librarianship and information studies for many years, I attempted to provide sufficient familiarity with statistical methods to enable them to understand the utilization of such methods when referred to in specialized areas of study such as management of libraries and bibliometrics. Rule-of-thumb procedures for performing typical calculations provided patterns to be followed and applied to other similar problems. It was hoped that mastery of such procedures would help students to show that information collected in projects could be manipulated in order to lead to useful conclusions and assist them to present information and results more effectively. The basis of that teaching is contained in my book *Basic statistics for librarians* (London, Library Association Publishing, 3rd ed., 1988).

The aims and objectives outlined above were achieved with a high degree of success even among students who claimed lack of numeracy and inability to understand mathematics. However, many students commented that they felt confident in performing operations and calculations on given data only if told what type of operation or calculation was involved. Without such instruction, they felt at a loss as to where to begin and indicated that there was a need for guidance on *what* to do with data presented to them as opposed to *how* to do it.

The aim of the present text is therefore to try to deal with the question of what to do with statistical data. I hope that, by using this short work in conjunction with my above-mentioned text, students will be able to deal appropriately and confidently at a basic level with any collection of data they encounter.

I am conscious of the growing demands on the library and information professions at large to handle data in their everyday work. I hope that they, as well as all students of librarianship and information management, will find this text of value.

ACKNOWLEDGEMENTS

Throughout this booklet I have used examples to help to clarify the points I make. Some of the collections of data are hypothetical but most have been gleaned over a period of many years. References to the sources of these data were not recorded but I gladly acknowledge to the authors, whoever they may be, their value in providing more realistic problems to illustrate the subject matter of the text.

I am particularly indebted to students attending Newcastle upon Tyne Polytechnic for their helpful comments and to my wife Mary for her encouragement.

INTRODUCTION

There are a number of aspects of interpreting statistical data which may be of interest. At the outset, it must be appreciated that there are two main categories of data — quantitative and qualitative — and that the methods of handling the data depend on which category is being dealt with. Chapter 1 -- *The nature of data* — defines these two categories, indicates how they may be distinguished and how the data are collected.

The quantity of data collected may be considerable; in fact, the greater the quantity, the more reliable and valuable will be the statistics. It is necessary to be able to specify such quantities of data concisely, either orally or in writing. This problem is dealt with in Chapter 2 — *Representation of data.*

Another aspect of communicating data is its presentation in the form of tables or diagrams. The choice between the various forms of presentation is discussed in Chapter 3 — *Presentation of data.*

Such collections of data will be of some interest in themselves, but they may well have been collected in order to draw conclusions regarding a particular problem which is under investigation. An appropriate statistical test or other manipulation of the data is then required. The various possibilities are dealt with in Chapter 4 — *Manipulation of data.*

As indicated in the Preface, the aim of this book is to give guidance on what to do with statistical data. In only a few instances are details given on how to perform calculations and draw diagrams. If you require the latter information, you could usefully consult *Basic statistics for librarians* (BSL 3rd ed.) — the bracketed notes in this book refer to page numbers in that work. Some examples of the various diagrams referred to are illustrated in the Appendix (pp.57−68).

Throughout each chapter, examples of data are given to illustrate the points made to help the reader to identify the procedures appropriate to any collection of data with which he or she may be concerned.

The following questions and answers to determine an operation of interest will provide a guide to which part of the book you should now turn.

Question 1: Do you know what type of data you are dealing with?

    Answer: Yes − Go to Question 2

               No − Read Chapter 1

Question 2: Do you wish to represent a large amount of data concisely?

    Answer: Yes − Read Chapter 2

               No − Go to Question 3

Question 3: Do you wish to present the data diagrammatically?

    Answer: Yes − Read Chapter 3

               No − Go to Question 4

Question 4: Do you wish to manipulate the data to draw conclusions?

    Answer: Yes − Read Chapter 4

               No − Go to Question 5

Question 5: Do you want more information?

    Answer: Yes − Try *Further reading* (pp.69−71)

               No − Sorry, no further options

Chapter 1

THE NATURE OF DATA

1.0 Data

Data consist of observations of some attribute which varies from one observation to another. For example, documents have the attribute of 'type of publication'; an online search has the attribute of 'time taken to undertake it'; a loans department has the attribute of 'number of issues recorded in a given period of time'. Clearly, in each of the examples given, the attribute varies from one observation to another: a document may be a report, a periodical, a textbook, etc.; different searches take different lengths of time to complete; different loans departments will make different numbers of issues in the given period of time. The attribute observed varies from one observation to another and therefore is referred to as a *variable*.

1.1 Collection of data

The data collected when observations are made are referred to as *raw data*. A *tally* is made of each observation in the form of a stroke and, to ease counting, the tally marks are usually grouped in fives as illustrated in Figure 1.1.

$$\text{卌 卌 ||}$$

1.1 Tally marks

Tally marks are recorded in a table known as a *tally chart* under the general headings of:

    attribute   *tally*   *frequency (of observation)*

Examples of tally charts can be seen in Figures 1.2 and 1.3.

1.2 Variables

There are two types of variable, *quantitative variables* and *qualitative variables*. Some observations may involve a single variable of either type; other observations may involve a combination of variables.

1.2.1 Single variables
1.2.1.1 Quantitative variables

A quantitative variable *involves a counting operation at each observation*. For example, the number of issues from loans departments is a quantitative variable because, for each department observed, the number of issues in a specified period of time is counted. Similarly, the number of books issued daily from any one library is a quantitative variable because, for each day observed, the number of issues is counted.

Another example of a quantitative variable would be the number of periodicals regularly read by readers. The number is likely to vary from one reader to another. As each reader is observed, the number of periodicals read is counted. The raw data could be arranged and tabulated in a tally chart as shown in Figure 1.2, a tally mark being recorded for each reader against the number of periodicals read. In the final column, the total number of tally marks in each row is given.

1.2 Tally chart of quantitative data

Attribute	Tally	Frequency
Number of periodicals read		Number of readers
0	‖‖	5
1	‖‖ ‖‖ ‖	12
2	‖‖ ‖‖ ‖‖ ‖	17
3	‖‖ ‖‖ ‖‖	15
4	‖‖ ‖‖ ‖	12
5	‖‖ ‖	7
6	‖‖	5
7	‖	2
	Total	75

So, when looking at a collection of data, ask the question: 'Was anything counted when each observation was made?' If your answer is 'yes', the variable is quantitative.

2

Example 1
In a survey of online searching operations, the connect-hours per file per month were noted as:

0.458	0.230	0.963	0.134	0.135	0.125
0.267	0.267	0.135	0.235	0.125	0.518

Clearly the survey involved keeping records over a period of a month of the usage of twelve different files. For each file observed, the number of connect hours were counted. Therefore the data are quantitative.

Example 2
In a small investigation to try to discover whether the inclusion of colour plates may influence the price of a paperback book, the prices of some illustrated paperback books (in pounds) were noted as:

1.25	0.55	1.05	0.50	0.75	0.85

These data were compared with the average price of non-illustrated paperback books which was £0.50.

In this case, when each book was observed, the number of pence required to buy it were counted. Therefore, again, the data are quantitative.

1.2.1.2   Qualitative variables
In contrast to a quantitative variable, each observation of a qualitative variable does not involve a counting operation but merely *an allocation to a category*. For example, the type of publication accessioned by a library is a qualitative variable − when each publication is observed, nothing is counted, the publication is merely categorized according to whether it is a report, a periodical, a textbook, etc. It is only after all observations have been made that anything is counted. Then the number of observations in each category are counted. In the example above, the total number of reports, total number of periodicals, total number of textbooks, etc., are obtained. The raw data can again be recorded in a tally chart, with a tally mark being made for each document against the appropriate attribute. Such a chart is illustrated in Figure 1.3.

So, when looking at a collection of data, if you ask the question: 'Was anything counted when each observation was made?', and the answer is 'no', the variable is qualitative. This conclusion would be confirmed by affirmative answers to further questions: 'When each observation was made was the action simply an allocation to a category?' and 'Do the numerical data merely consist of the total number of observations allocated to each category?'

3

1.3 Tally chart of qualitative data

Attribute	Tally	Frequency
Type of document		Number of documents
Periodical	JHT JHT JHT JHT	20
Report	JHT JHT IIII	14
Textbook	JHT III	8
	Total	42

Example 3

In a survey of sources of books read by teenagers, it was found that 11 came from the school library, 2 from the public library, 4 from friends, 6 from the family and 13 were bought.

The data were obtained from a group of teenagers, each one being asked from where they obtained the books they had read. The actual observations related to individual books and when each book was considered, nothing was counted. The book was allocated to one of the five categories and when all the observations had been made, the books in each category were counted. Therefore the data are qualitative.

Example 4

In a survey of career aspirations of library-school students, 8 aspired to be a county librarian, 10 a divisional librarian, 10 the head of a reference library, 10 the head of a schools' service, 7 a branch librarian and 4 a children's librarian.

Here, 49 students were observed and each allocated to one of the six categories. Nothing was counted until all students had been questioned and the number in each category was found. Therefore, again, the data are qualitative.

1.2.2 Aspects of interest

It should be noted that some items of interest have more than one attribute, some of which may be quantitative and some qualitative. For example, in examining a volume of abstracts, several attributes of each abstract may be observed. For example:

1 the *length* of abstract may be of interest, in which case a *quantitative* observation is made by counting the number of words or lines in each abstract observed.

4

2   the *time lag* between publication of the primary document and publication of the abstract may be of interest, in which case *quantitative* observations are made by counting the number of months between the date recorded in each abstract and the date of the volume of abstracts.

3   the *country of origin* of items abstracted may be of interest, in which case a *qualitative* observation is made by allocating each abstract to the country indicated.

4   the *type of document abstracted* may be of interest, in which case a *qualitative* observation is made by allocating each abstract to the document type indicated.

1.2.3   Combinations of variables

1.2.3.1   Qualitative/quantitative variables

In some instances observations involve two variables, one of which is qualitative and the other quantitative, the qualitative variable often being time or date. For example, a number of volumes of an abstracts publication may be examined. As each volume is observed, it can be allocated to the qualitative category of year of publication and the quantitative number of abstracts in the volume can be counted.

Other examples of a similar type would be variation over the years of loan charges for special materials, of document issues, of book cost index, etc.

You may suspect that data fall into this category when there is *a series of time categories*. Note then whether the observation for each time category involves *a counting operation*.

Example 5

Books published in Britain have been recorded as follows:

Year	Books published
1950	17072
1955	19962
1960	23783
1965	26358
1970	33489
1975	35608
1980	48157

The *time categories* of date are given in the first column and, for each year, the *number* of books published was counted.

Example 6

The daily issues of junior non-fiction from a library during the course of a week were recorded as follows:

Day	Number of issues
Monday	39
Tuesday	14
Wednesday	21
Thursday	47
Friday	36
Saturday	96

Here the time categories are the days of the week and, for each day observed, the number of issues were counted.

1.2.3.2 Quantitative/quantitative variables

In yet other instances, observations may involve two variables which may be related to each other, both of which are quantitative. For example, when each of a collection of books is observed, the number of pages it contains can be counted and the number of pounds required to buy it can also be counted. In another example, for each library user observed, the number of miles travelled to use the library can be counted and the number of visits made in a given period of time can also be counted.

Each observation is, in fact, a double observation of apparently related variables and each observation involves *two counting operations*.

Example 7

In an investigation of the efficiency of the information retrieval process, the following data were obtained:

Search number	Total documents retrieved	Relevant documents retrieved
1	79	21
2	18	10
3	20	11
4	123	48
5	16	8
6	109	48
7	48	25
8	2	1
9	36	5

Each search was examined by counting the total number of documents retrieved and by counting the number of those documents which were relevant to the specific interest. So both variables – which may be expected to be related – are quantitative. The 'search number' is not a variable but merely an identification number for each search observed.

Example 8
Figures relating to document usage were recorded as follows:

Document number	Age of document (years)	Frequency of use (times/year)
1	1	40
2	3	18
3	2	30
4	4	21
5	3	26
6	5	10
7	4	13
8	3	35

When each document was observed, two counts were made, *viz*: the number of years since it was published and the number of times it had been used in the previous year. There are thus two, probably related, quantitative variables.

1.3 Data handling
The methods of representing, presenting and manipulating data depend on the nature of the data involved and therefore, before proceeding further, it is essential that the reader should be able to determine to which of the four categories described above a particular collection of data belongs. The exercises at the end of the chapter provide an opportunity to test the understanding of this.

When you feel confident in determining the nature of a set of data, the following questions and answers will indicate where to read next.

Question 1: Are the data *quantitative*?

Answer: Yes – Go to Question 1.1

No – Go to Question 2

Question 1.1: Do you wish to represent the data?

Answer: Yes – Read Chapter 2, Sections 2.0 and 2.1

No – Go to Question 1.2

Question 1.2: Do you wish to present the data?
  Answer: Yes − Read Chapter 3, Sections 3.0 and 3.1
           No − Go to Question 1.3
Question 1.3: Do you wish to manipulate the data?
  Answer: Yes − Read Chapter 4, Sections 4.0 and 4.1
           No − Sorry, no further options

Question 2: Are the data *qualitative*?
  Answer: Yes − Go to Question 2.1
           No − Go to Question 3
Question 2.1: Do you wish to represent the data?
  Answer: Yes − Read Chapter 2, Sections 2.0 and 2.2
           No − Go to Question 2.2
Question 2.2: Do you wish to present the data?
  Answer: Yes − Read Chapter 3, Sections 3.0 and 3.2
           No − Go to Question 2.3
Question 2.3: Do you wish to manipulate the data?
  Answer: Yes − Read Chapter 4, Sections 4.0 and 4.2
           No − Sorry, no further options

Question 3: Are the data qualitative/quantitative?
  Answer: Yes − Go to Question 3.1
           No − Go to Question 4
Question 3.1: Do you wish to represent the data?
  Answer: Yes − Read Chapter 2, Sections 2.0 and 2.3
           No − Go to Question 3.2
Question 3.2: Do you wish to present the data?
  Answer: Yes − Read Chapter 3, Sections 3.0 and 3.3
           No − Go to Question 3.3
Question 3.3: Do you wish to manipulate the data?
  Answer: Yes − Read Chapter 4, Sections 4.0 and 4.3
           No − Sorry, no further options

Question 4: Are the data quantitative/quantitative?
  Answer: Yes − Go to Question 4.1
           No − No further options; re-read Chapter 1
Question 4.1: Do you wish to represent the data?
  Answer: Yes − Read Chapter 2, Sections 2.0 and 2.4
           No − Go to Question 4.2
Question 4.2: Do you wish to present the data?
  Answer: Yes − Read Chapter 3, Sections 3.0 and 3.4
           No − Go to Question 4.3

Question 4.3: Do you wish to manipulate the data?
  Answer: Yes — Read Chapter 4, Sections 4.0 and 4.4
             No — Sorry, no further options

Exercises

Exercise 1
Are the following variables quantitative or qualitative?
1.1   The height of a book
1.2   The subject of a book
1.3   The purpose for which a xerox copy is made
1.4   The price of a book
1.5   The allocation of a library budget
1.6   The size of population served by a library
1.7   The place of residence of a library user
1.8   The size of a bibliography
1.9   The methods of information transfer employed by library users
1.10 The sex of authors

Exercise 2
Identify the variables and the variable types involved in each of the following examples.
2.1   Age and frequency of use of a document
2.2   Live stock and daily issues from a library
2.3   Monthly figures for issues of fiction by a library
2.4   Annual expenditure on staff and stock by a library
2.5   Items obtained and supplied by a library in a specified time
2.6   Growth of scientific periodicals
2.7   Fiction and non-fiction issues from a library in a given length of time
2.8   Fiction issues from a library over a period of years
2.9   The cost and content of a volume of abstracts
2.10 Decrease in online search time due to experience gained by searcher

Exercise 3
What types of data are involved in the following examples?
3.1 Statistics for a number of Metropolitan City Libraries included the following data:

Population ('000 000s)	Stock ('000 000s)
3.31	3.54
2.83	4.32
2.60	2.94
2.45	2.14
2.22	0.79
1.95	2.79
1.74	1.65
1.61	0.62
1.51	2.27
1.43	1.54
1.32	1.50

3.2 In a survey of library users, their occupations were noted as follows:

Occupation	Number of users
Employed	576
Housewife	599
Pupil	215
Student	43
Unemployed	156
Retired	600

3.3 A study of the growth of the literature on liquid crystals provided the following data:

Year	1965	1966	1967	1968	1969	1970	1971	1972	1973	1974
Reviews	1	3	2	8	18	12	20	29	52	72
Patents	0	2	1	2	2	10	22	47	144	210

3.4 In an attempt to estimate the size of field required in a computer store to accommodate a list of stopwords for compiling an index, the following data were collected:

No. of characters/stopword	3	4	5	6	7	8	9	10	11
No. of stopwords	23	38	33	36	16	3	4	3	1

Chapter 2

REPRESENTATION OF DATA

2.0 Introduction

If data are to be useful and produce reliable statistics, the greater the quantity of data the better. However, large quantities of data are not easy to assimilate or to communicate. Look, for example, at the data in Figure 2.1 relating to mobile library services.

15	5	16	10	9	16	13
13	10	20	15	9	13	16
15	20	13	16	9	15	14
10	15	9	16	13	15	14

2.1 The number of stops made by 28 mobile library services

Anyone interested in statistics of such services could be given that collection of figures. It would provide a full, detailed and accurate picture of the number and variability of stops made by mobile library services. However, even for a comparatively small collection of data such as this, it is not a convenient way of conveying information. Minute detail is often not essential. If you were asked to provide information on the height of human beings, a detailed listing of the individual heights of the world's population is probably not required — it may be sufficient to supply the information that the average height is, say, 170 centimetres. Similarly, in the case of the mobile library services, it may be sufficient to know the average number of stops.

As these two examples demonstrate, we often want to represent a mass of data by a single figure. Precisely how a mass of data is represented depends on the nature of the data. In this chapter we will look at methods of representing the various types of data defined in Chapter 1.

11

2.1 Quantitative data

Measures of average are the *mode*, the *mean* and the *median*.

In addition, **measures of dispersion** may be necessary, *viz: range, standard deviation* and *quartiles*.

A further measure of interest is *skewness*.

2.1.1 *Mode* (BSL, 3rd ed., 15 – 16, 19)

The mobile library service data in Figure 2.1 are quantitative since, for each service observed, the number of stops were counted. As in the example concerning the heights of human beings, it may be quite sufficient to represent all 28 observations by a single figure. There are several ways in which such a single figure can be found, the easiest being to select that value of *the variable quantity which occurs most frequently*. That value is the *modal value* – or the *mode*. For the data in Figure 2.1, 15 stops occurs more often than any other number of stops (six times); therefore the modal number of stops is 15.

The mode is easy to find and it is not affected by extreme values. For example, even if one service made as many as 40 stops, the modal value of 15 would not have been affected. However, the mode is not a very reliable measure of average. In our example, services with 13 stops and 16 stops were both observed five times, so these were very close to being the modal value. For these reasons the mode cannot be used for subsequent statistical calculations.

Nevertheless, it can be useful as a quickly-obtained measure of the value of the variable most likely to be observed, especially for comparative purposes. For example, the pattern of time-lag (i.e. the lapse of time between the primary publication and publication of its abstract) is probably similar for different abstracting services, in which case quoting their respective modal time-lags will provide an indication of which is likely to be the most up-to-date. An abstract bulletin with a modal time-lag of three months will give more up-to-date information than a bulletin with a modal time-lag of six months. That is not to say that some abstracts in the latter volume might not have a time-lag of less than three months – the modal value of six months only indicates that that is the most commonly occurring value.

2.1.2 *Mean* (BSL, 3rd ed., 20 – 3)

A measure of the average which is representative of all the observed values of the variable and consequently of more use for performing subsequent statistical calculations is the *mean*. This is found by adding

12

together all the observed values of the variable quantity and dividing the result by the number of observations. This is the measure usually meant by the common use of the word 'average'.

The mean number of stops made by the mobile library services noted in Figure 2.1 can be found to be approximately 13.36. This figure gives the reader an immediate idea of the number of stops if all services were the same. However, it is unlikely that all services will be the same. Some may have more than 13.36 stops, others may have fewer — as, indeed, a glance at the detailed data in Figure 2.1 shows.

2.1.3 Dispersion

It could be important to know *how much the data vary*.

In the mobile library example, does the number of stops actually vary only a little from service to service (e.g. between 12 and 14) or does it vary considerably (e.g. between 3 and 23)? Let us suppose that each stop corresponds to half an hour of time. If a service makes the mean number of 13.36 stops, it would quite comfortably complete its itinerary in a seven hour day. Similarly, if the number of actual stops differs only a little from the mean (between 12 and 14 stops), the itinerary will take only six or seven hours. However, a look at the raw data shows that some services have as many as 20 stops which, at half an hour per stop, would take ten hours and could not be completed in a single working day. At the other extreme, the service with only five stops would have a great deal of time to spare. The assumption of half-hour intervals could of course be incorrect!

To give a better description of the data therefore, a measure of the spread — or *dispersion* — of the observations needs to be given, in addition to the mean.

2.1.3.1 Range (BSL, 3rd ed., 26)

The easiest way of defining the dispersion is to quote the *range*. This is simply *the difference between the smallest and largest observations*. In the mobile library example the range is 15 (20−5).

This figure gives an idea of variability, but it can be misleading if the extreme value or values are isolated exceptions. In our example the smallest observation (5) is well below the main group which starts at 9, whilst the two observations of the largest value (20) are well above the main group which finishes at 16. A more reliable figure for the range in this case would therefore be 7 (16−9) but this is not what is derived if we follow the definition of range.

13

2.1.3.2   Standard deviation (BSL, 3rd ed., 28−30)
A better way of indicating the degree of dispersion is to quote the *standard deviation*, which for our example can be calculated from the raw data (Figure 2.1) to be approximately 3.38.

About 68% of data fall within one standard deviation of the mean so, by coupling the standard deviation with the mean, we can state that roughly 68% of all services theoretically make 13.36±3.38 stops, i.e. between 9.98 (13.36−3.38) and 16.74 (13.36+3.38) stops.

Also about 95% of the data fall within two standard deviations of the mean so we can go on to say that roughly 95% of all services theoretically make 13.36±(2 × 3.38) stops, i.e. between 6.6 and 20.12 stops.

2.1.4   Skewness about the mode (BSL, 3rd ed., 34)
In practice it is usually found that the *observations are not equally spread on either side of the mean.*

If there are more observations with a smaller value than the mean than there are observations with a larger value, the data are said to have a *positive skewness.*

If there are more observations with a larger value than the mean than there are observations with a smaller value, the data are said to have a *negative skewness.*

In the case of the mobile library services, 13 made less than 13.36 stops and 15 made more. There is some degree of negative skewness in this set of data.

A common measure of skewness is given in Figure 2.2.

$$\text{Skewness} = \frac{\text{Mean} - \text{Mode}}{\text{Standard deviation}}$$

2.2   Formula for calculating skewness using mode

In our example, the mean is 13.36, the mode 15 and the standard deviation 3.38. Using the formula in Figure 2.2 the skewness can be calculated as follows:

$$\text{Skewness} = \frac{13.36 - 15}{3.38}$$

$$= -0.485$$

This shows a moderate degree of negative skewness; a value of −1 (or +1 if the skewness is positive) indicates a fairly high degree of skewness.

14

2.1.5 Mean, dispersion and skewness

In order to represent a collection of quantitative data concisely, it is therefore necessary, and usually sufficient, to quote three pieces of information, *viz*:

1 the value of the mean;
2 the value of the standard deviation;
3 the nature of the skewness.

This method of representing data is the most generally useful since it is based on all the data and consequently can be used for further statistical calculations. However, there are occasions when discretion must be exercised. Look, for example, at the data on book prices in Figure 2.3.

20	27	21	35	17
25	27	30	32	200
32	35	20	35	24

2.3 Cost (£) of a selection of books

The mean cost would be found to be £38.67. However, to use this to estimate future expenditure on books would be misleading since it is based on a calculation including an encyclopedia costing £200. The mean so calculated is therefore unduly inflated. A more satisfactory value of the mean would be obtained from the 14 observations left if the encyclopedia is omitted, *viz* £27.14.

2.1.6 Median (BSL, 3rd ed., 16–17, 19–20)

In the last example, the effect of the unusually high cost of the encyclopedia could be nullified by using the *median* as a measure of the average value.

To find the median, the data must first be put in an *array* in *ascending order of magnitude* as in Figure 2.4

17 20 20 21 24 25 27 27 30 32 32 35 35 35 200

2.4 Cost (£) of a selection of books in an array

The array is then divided into two equal parts and the value of the *middle observation* (or the mean of the two observations on either side of the middle of an array with an even number of data) is noted as the median. In this example, the median cost of a book is £27. There are as many observations of £27 or less as there are observations of £27 or more. The extreme observation of £200 does not affect the median value.

2.1.7 *Quartiles* (BSL, 3rd ed., 31–3)

When the median is used to represent the average value of a collection of data, the dispersion is measured by *quartile* values.

The *lower quartile* is that value in the array below which there are one quarter of the observations. In the current example (Figure 2.4), the lower quartile is £21.

The *upper quartile* is that value in the array below which there are three quarters of the observations. In the example (Figure 2.4), the upper quartile is £35.

The difference between the quartiles (in the example, £14 (35−21)) is the *interquartile range*. This is a measure of the range of the variable quantity containing half of the observations.

With larger quantities of data, *deciles* may also be used − the *first decile* being that value of the variable below which there are one tenth of the observations and the *ninth decile* being that value of the variable below which there are nine tenths of the observations.

The quartiles and the median divide the array into four equal groups of observations, the deciles into ten equal groups ... and *percentiles* into one hundred equal groups.

Such representation can be particularly useful when looking at certain special collections of data. For example, when considering data on salaries, if the mode is used as the average, the information conveyed is merely that salary which is most frequently observed. The modal value is not affected by the sizes of salary which are observed less frequently. If the mean is used as the average, although all observed salary levels are included in the calculation, the average value would be inflated by any exceptionally high salaries included in the survey. However, if the median is used as the average, it will not be affected by any exceptional salaries. It will divide the array of observations into two equal groups, half with a salary equal to or less than the median, half with a salary equal to or greater than the median. That will satisfy a reader's interest in whether they are in the bottom half or the top half of the salary range. Also quoting the quartiles will indicate whether they are in the middle range, among the poorly paid or among the high fliers.

2.1.8 *Skewness about the median* (BSL, 3rd ed., 35)

Whilst the median divides the array of variables into two equal groups,

16

in practice the values of the variable above the median may be more widely spread than those below, showing positive skewness. Figure 2.5 shows such a set of data, with a range below the median of 3 (10−7) and a range above the median of 8 (18−10).

| 7 | 8 | 8 | 9 | 9 | 10 | 12 | 14 | 15 | 16 | 18 |

2.5 An array of salaries (in thousands of pounds) with a median of £10 000, showing positive skewness

If the values of the variable are more widely spread below the median than above, as in Figure 2.6, the skewness is negative.

| 4 | 6 | 7 | 8 | 9 | 10 | 11 | 11 | 11 | 12 | 13 |

2.6 An array of salaries (in thousands of pounds) with a median of £10 000, showing negative skewness

A formula for calculating the skewness when the median is used as the average is given in Figure 2.7

$$\text{Skewness} = \frac{3 \times (\text{Mean} - \text{Median})}{\text{Standard deviation}}$$

2.7 Formula for calculating skewness using median

2.1.9 Mode, mean and median

For the data given in Figure 2.6, one can find the following statistics:

Mode	11
Mean	9.27
Range	9
Standard deviation	2.64
Skewness (using Figure 2.2)	−0.66
Median	10
Lower quartile	7
Upper quartile	11
Interquartile range	4
Skewness (from Figure 2.7)	−0.83

2.1.10 Grouping of data

In the examples used so far in this chapter, the variable quantities have assumed only a small number of different values. For example, in

Figure 2.1 there are only 16 possible values between 5 and 20 and in Figure 2.6 there are only 10 possible values between 4 and 13. On many occasions the number of possible values is very much greater, as in Figure 2.8.

19.95	10.50	25.00	30.00	30.00	15.00	25.00
9.95	9.95	30.00	30.00	12.00	14.50	25.00
27.50	30.00	16.00	30.00	22.50	27.50	18.00
12.50	32.50	18.00	17.50	32.50	21.00	45.00
21.00	9.95	35.00	25.00	17.00	21.00	12.50
35.00	32.50	12.00	9.50	32.50	17.00	30.00
15.00	35.00	15.00	9.95	27.50	35.00	21.50
17.50	20.00	22.00	17.50	28.00	7.50	20.00
15.00	35.00	45.00	19.50	30.00	21.00	16.00
12.00	24.00	7.50	18.50	24.00	12.00	30.00
25.00	27.50	45.00	17.00	55.00	37.50	3.50
8.00	37.50	40.00	20.50	12.00	24.00	12.50
5.95	30.00	15.00	18.00	55.00	35.00	40.00
27.50	15.00	20.00	20.00	25.00	22.50	27.50

2.8 Prices (£.p) of volumes advertised in a publisher's list

Such data are far easier to handle if put into groups, as in Figure 2.9.

2.9 Prices (£.p) of volumes advertised in a publisher's list (grouped)

Cost (£.p)	Number of volumes
0.00 – 4.99	1
5.00 – 9.99	9
10.00 – 14.99	10
15.00 – 19.99	20
20.00 – 24.99	16
25.00 – 29.99	13
30.00 – 34.99	14
35.00 – 39.99	8
40.00 – 44.99	2
45.00 – 49.99	3
50.00 – 54.99	0
55.00 – 59.99	2
	Total 98

Measures of average and dispersion can be derived from the grouped data in much the same way as before, to describe the collection of data. The mode may be given as the *modal class*. In this example the modal class is £15.00–£19.99.

Alternatively, the mode (an individual value) may be calculated from the data using the following formula:

$$\text{Mode} = B + \left(\frac{D_1}{D_1 + D_2}\right) \times C$$

where B = lower boundary of the modal class

     D_1 = difference between number of observations in the modal class and the number of observations in the next lower class

     D_2 = difference between number of observations in the modal class and the number of observations in the next higher class

     C = width of the modal class

In this case,

$$\text{Mode} = 15 + \left[\frac{20-10}{(20-10)+(20-16)}\right] \times 5$$
$$= 18.57$$

The mean can be calculated as follows:

$$\text{Mean} = \frac{\text{sum of (mid-point of each interval} \times \text{no. of observations in each interval)}}{\text{total no. of observations}}$$

In this case:

$$\text{Mean} = \frac{(2.5 \times 1)+(7.5 \times 9)+(12.5 \times 10)+\ldots+(57.5 \times 2)}{98}$$
$$= 24.08$$

The standard deviation is 11.2 and the skewness is +0.49 (using the Figure 2.2 formula).

Lastly the median can be calculated from the data using the following formula:

$$\text{Median} = B_1 + \frac{x}{y} \times C_1$$

where B_1 = lower boundary of the class in which the middle observation of the array lies

     x = number of observations to be added to the cumulative total in the previous classes in order to reach the middle observation in the array y = number of observations in the class in which the middle observation of the array lies

C_1 = width of the class in which the middle observation of the array lies

In this case,

$$\text{Median} = 19.995 + \frac{9}{16} \times 5$$
$$= 22.81$$

The quartiles are approximately 16.10 and 31.60, the interquartile range 15.50 and the skewness +0.34 (using the Figure 2.7 fomula). From these figures, one can state the following:

1   The most common cost of a book was in the range £15.00 to £19.99, theoretically approximately £18.50. Also, there was a greater spread of prices above £18.50 than below £18.50.

2   Half the books cost £22.80 or less and half cost £22.80 or more, whilst half of them cost between £16.10 and £31.60. The spread of prices above £22.80 was greater than the spread of prices below £22.80.

3   The average price of all books advertised is £24.08, from which a budget for say 100 books can be estimated to be £2408.

4   From the standard deviation it can be deduced that 95% of the books cost between £1.64 and £46.52.

2.1.11 Samples and populations (BSL, 3rd ed., 44–8)

When collecting data, it may not be practicable to make observations on a whole *population*. In a statistical context 'population' does not necessarily refer only to persons. It refers to the whole group of whatever is being investigated which may be persons or may be, for example, book issues, online searches, abstracts, mobile library services or bibliographies.

If the population is large, the processes of observation and calculation would be very long and tedious, if not impossible. In such circumstances, it is more economical to make use of only a part of the population. That part which is being investigated is known as the *sample*. However, it is the population which is of real interest, so any observations made on a sample need to be utilized to obtain information regarding the population.

Let us assume now that the data in Figure 2.1 represent only a sample

20

of mobile library services. We deduced that, for this sample, the mean number of stops was 13.36 and the standard deviation was 3.38. We now wish to deduce what the mean and the standard deviation might be for the whole population of services, from which the sample was selected at random.

There are three ways of tackling the problem:

1 The easiest way is to consider the results obtained from the sample as a *point estimate* and assume that the mean and standard deviation of the population will be the same as those of the sample, *i.e.* 13.36 and 3.38, respectively.

The larger the sample, the more reliable a point estimate is likely to be as a measure of the parameters of the population. However, we have no idea of the accuracy of such a conclusion.

2 In order to obtain a measure of the accuracy of an estimate, use is made of the *standard error of the mean*. This figure is found by dividing the standard deviation by the square root of the number of observations in the sample:

$$\text{Standard error} = \frac{\text{standard deviation}}{\sqrt{\text{sample size}}}$$

In this case:

$$\text{Standard error} = 3.38 \div \sqrt{28}$$
$$= 0.64$$

By adding twice the standard error to the mean of the sample, an upper limit to the estimate of the mean of the population is obtained:

$$13.36 + (2 \times 0.64) = 14.64$$

Similarly, the lower limit to the estimate is found by subtracting twice the standard error:

$$13.36 - (2 \times 0.64) = 12.08$$

From the theory on which this calculation is based, it can be said with a high degree of confidence that the mean of the total mobile library service population will lie between 12.08 and 14.64. It can also be shown that, for large samples, the standard deviation of the population will be approximately the same as that for the sample, *i.e.* 3.38.

Although this calculation involves a minimum amount of effort since it is based on one sample only and although the confidence in the result is high, it is very dependent on the quality of the one sample on which the calculation is based.

3 In order to obtain a measure of accuracy without being dependent on the quality of a single sample, several samples must be used. Then, if one of the samples should be unsatisfactory, its effects on the overall result will be reduced. The mean and standard deviation and standard error must be calculated for each sample. The statistics in Figure 2.10 were obtained from the data in Figure 2.1 (sample 1) and two other similar samples.

2.10 Statistics relating to three samples of mobile library services

Sample number	Mean	Standard deviation	Standard error
1	13.36	3.38	0.64
2	13.21	3.23	0.61
3	13.39	3.18	0.60

From the values in Figure 2.10, the following information can be deduced:

(a) *Mean value of means* $= \dfrac{\text{sum of means of samples}}{\text{number of samples}}$

$= (13.36 + 13.21 + 13.39) \div 3$

$= 13.32$

(b) *Mean value of standard error* $= \dfrac{\text{sum of standard errors of samples}}{\text{number of samples}}$

$= (0.64 + 0.61 + 0.60) \div 3$

$= 0.62$

(c) The *population mean* probably lies between the mean value of the means plus/minus the mean standard error, i.e. between 12.70 $(13.32 - 0.62)$ and 13.94 $(13.32 + 0.62)$.

Although it takes longer to obtain the final result using this procedure, one can have greater confidence in calculations based on several samples. Of course, the greater the number of samples, the greater the confidence but also the greater the time and effort required.

Example 9

The number of junior books issued from a public library on a random sample of days were:

54   47   17   70   18   58   38   38   31   50   54   27

A further five samples were taken, each including observations on 12 days, with the following (calculated) results:

Sample number	Sample mean	Sample standard deviation
2	41.67	16.08
3	43.67	17.82
4	39.50	21.14
5	36.08	20.06
6	45.50	20.40

The mean number of daily issues of junior books can be estimated from the data in the first sample. The accuracy of that estimate can be determined by using the second procedure described in Section 2.1.11.

Using the results for samples 2 to 6, the maximum value which may be expected for the mean number of daily issues as determined by the standard error can be found by using the third procedure described in Section 2.1.11. Remember that the standard error is found by dividing the standard deviation by the square root of the number of observations in the sample (in this case 12 for each sample), *not* the number of samples.

Having estimated the maximum value of the mean number of daily issues by either of these methods, the maximum number of issues which may be expected in a year can be found by multiplying that value of the mean by the number of working days in the year.

2.2 Qualitative data

The variable in qualitative data is a category, not a number. Because of this essential difference, qualitative data cannot be represented in the same way as quantitative data. In fact, the very nature of qualitative data leads to simple representation in the form of totals for the various categories involved.

In contrast to the 98 observations of the quantitative variable shown in Figure 2.9, Figure 2.11 shows a collection of 98 observations of a qualitative variable. In 1978−9, an investigation was made of the knowledge about British Library MARC records from the point of view of their inclusion of ERIC records. There were only three categories of response, *viz*: 'True', 'False' and 'No response', as shown in Figure 2.11.

2.11 Responses in investigation of British Library MARC records

True	24
False	44
No response	30
Total	98

It will be apparent from a comparison of Figures 2.9 and 2.11 that the representation of the two types of data present very different problems.

2.3 Qualitative/quantitative data

As indicated in Section 1.2.3.1 (page 5), this type of data is commonly concerned with the change in value of a quantitative variable over a period of time. There are two aspects which may be of interest:

1 an *index* to measure the change in a variable over a period of time;
2 a *time series analysis* to represent the variation throughout the time period.

2.3.1 Indexes (Indices)

An index is used to compare the value of a commodity or collection of commodities at the end of a time period with the value at the beginning of that time period. The value of the commodity at the beginning of the period is taken to be 100 and the value of the commodity at the end of the period is the index. There are various types of index, the choice of which depends on the sort of comparison required.

2.3.1.1 Simple index (BSL, 3rd ed., 124–6)

If the value of the commodity in any *given year* is to be compared with the value in the same *base year*, a *simple aggregative index* is used:

$$\text{Simple aggregative index} = \frac{\text{value in given year}}{\text{value in base year}} \times 100$$

Figure 2.12 shows some book issue statistics for a number of years.

2.12 Total number of issues of volumes of non-fiction by a library in a number of years

Year	1960	1961	1962	1963	1964
Number of issues	8094	9288	8416	9271	8233

If 1960 is taken as the base year, the simple aggregative index can be calculated to compare the number of issues for the year in question with the number of issues in 1960 (Figure 2.13).

2.13 Simple indexes for issues of volumes of non-fiction by a library in a number of years (1960 = 100)

Year	1960	1961	1962	1963	1964
Index	100	114.75	103.98	114.54	101.72

Thus, in 1962 the number of issues was 103.98 compared with 100 in 1960 and in 1963 the number of issues was 114.54 compared with 100 in 1960.

Indexes make it very easy to identify percentage changes. Between 1960 and 1962, there was a 3.98% increase in the number of issues; between 1960 and 1963, there was a 14.54% increase.

2.3.1.2 *Chain base index* (BSL, 3rd ed., 128–9)

If the value of the commodity in any *given year* is to be compared with the *previous year*, rather than with a base year, a *chain base index* is used. Figure 2.14 shows such an index for the data of Figure 2.12.

2.14 Chain base index for issues of volumes of non-fiction by a library in a number of years

Year	1961	1962	1963	1964
Index	114.75	90.61	110.16	88.80

This shows the number of issues in 1962 to be 90.61 compared with 100 in 1961 and the number in 1963 is shown to be 110.16 compared with 100 in 1962.

2.3.1.3 *Weighted index* (BSL, 3rd ed., 126–8)

A drawback of the simple index is that its value depends on the way in which a commodity is measured. For example, if we are concerned with the costs of publications, is the cost of a periodical taken as the cost of each issue or of its annual subscription?

The *weighted aggregative index* takes into consideration the quantities of each commodity involved in the calculation. As with the simple index, comparisons are made between given years and a single base year. However there is a choice between *base year weighting* and *given* (or *current*) *year weighting*.

25

An index with *base year weighting compares what it would cost* in the given year to buy the quantity of goods that actually were bought in the base year, *with what it did cost* to buy those goods in the base year.

$$\text{base year weighted index} = \frac{\text{price in given year} \times \text{quantity in } \textit{base} \text{ year}}{\text{price in base year} \times \text{quantity in base year}} 100$$

An index with *given year weighting compares what it costs* to buy a quantity of goods in the given year, *with what it would have cost* to buy the same quantity of goods in the base year.

$$\text{given year weighted index} = \frac{\text{price in given year} \times \text{quantity in given year}}{\text{price in base year} \times \text{quantity in } \textit{given} \text{ year}} 100$$

2.3.2 Time series analysis (BSL, 3rd ed., 133–9)

When a quantity is measured at regular intervals over a period of time, successive values are likely to vary appreciably and a graph of the data may appear very erratic. Nevertheless, it may be necessary to discover whether there is any *long-term trend* and the extent of any *short-term variation* in the data.

A *moving average* (BSL, 3rd ed., 135–7) smooths the data to enable information on both the trend and the variation to be derived.

Example 10

The following are daily issues of junior non-fiction from a library:

Day	Week 1	Week 2	Week 3
Mon.	36	46	66
Tues.	31	55	76
Wed.	25	37	40
Thurs.	55	80	74
Fri.	45	66	90
Sat.	90	115	150

The moving average will demonstrate whether, over the three week period, the general trend is an increase or decrease in daily issues. The *cyclical variation* is the difference between the moving average (expected) value and the corresponding actual daily figure. This will show the extent

26

to which the daily figures vary above and below the general trend. Figure 2.15 shows the result of such an analysis.

2.15 Time series analysis of data of Example 10

Week	Day	Number of issues	Moving average	Cyclical variation
1	M	36		
	T	31		
	W	25		
	T	55	47.9	+7.1
	F	45	50.7	−5.7
	S	90	53.7	+36.3
2	M	46	56.8	−10.8
	T	55	60.6	−5.6
	W	37	64.4	−27.4
	T	80	68.2	+11.8
	F	66	71.6	−5.6
	S	115	73.6	+41.4
3	M	66	73.3	−7.3
	T	76	74.8	+1.2
	W	40	79.8	−39.8
	T	74		
	F	90		
	S	150		

The moving average shows a general upward trend (*i.e.* increasing daily issues). The cyclical variation figures show that Saturday issues are markedly above the general trend and Wednesdays' are markedly below.

From this example, it may be appreciated how such information may be necessary/useful. If it is assumed that the number of issues is related to the number of staff required on the issue desk, the upward trend could forewarn that, if the trend continues, an extra permanent member of staff may be required on the issue desk. On the other hand, the cyclical variation suggests that, irrespective of the long-term requirement, an extra member of staff is probably needed on the issue desk on Saturdays, but perhaps one might be released on Wednesdays.

2.4 Quantitative/quantitative data
In Section 1.2.3.2 (page 6), it was indicated that, in this category of data, the two variables observed may be related to each other. For example, perhaps the greater number of pages in a book, the greater the cost. On the other hand, the greater the distance which must be travelled to visit a library, the fewer the number of visits which may be made in a given time period.

2.4.1 Correlation coefficients

To represent the degree of closeness of the connection between the two variables a *correlation coefficient* is quoted. If both variables increase together, the correlation coefficient is positive and if one variable increases while the other decreases, the correlation coefficient is negative. If there is no apparent relationship between the two variables, the correlation coefficient will be zero. The maximum value of the coefficient is $+1$ or -1 (according to whether the correlation between the variables is positive or negative).

Example 11

The following table gives the approximate number of abstracts (in thousands) in a selection of volumes of abstracts together with the cost of each volume:

Number of abstracts (thousands)	Cost (£)
36.7	115
8.5	52
12.5	75
3.9	31
0.5	9
1.3	12
4.1	20
19.4	56
4.3	24

The correlation coefficient determined from the data is $+0.938$ which suggests that there is a close relationship between the number of abstracts and the cost. However, if the number of observations is small, the closer must the correlation coefficient be to one before it is significant. An appropriate test should be carried out to determine whether or not a coefficient may be considered significant.

When actual values of the two variables are known and used in the calculation (as in Example 11), a figure known as the *Pearson correlation coefficient* can be derived (BSL, 3rd ed., 158–61). In some cases, the actual values of the variables are not known, only a rank order, as in Example 12. The method of calculation is different but a measure of relationship known as the *Spearman correlation coefficient* can be determined (BSL, 3rd ed., 161–3).

Example 12

Boys and girls were questioned about their reading interests and asked to put various types of novel into their order of preference, with the following results:

| Type of novel | Rank orders | |
	Boys	Girls
Animal stories	4	2
Historical novels	3	3
Romances	5	1
War stories	1	5
Westerns	2	4

The value of the Spearman Correlation Coefficient turns out to be -1 showing a perfect inverse relationship between the interests of the boys and the girls.

Exercises

Exercise 4

What types of data are given in the following examples and what type of figure(s) may be used to represent them?

4.1 The following table shows the number of times per year that each of a random sample of documents of different ages was used:

Age of document (years)	1	3	2	4	3	5	4	3
Frequency of use (times/year)	40	18	30	21	26	10	13	35

4.2 The original languages of translations of religious works in 1974−6 were:

Language	Number of works
French	70
German	101
Latin	52
Spanish	23

4.3 In a bibliography of the early development of mass spectrometry, the dates of references were noted. The number of items published in each year were:

Year	Number of items	Year	Number of items
1938	32	1948	109
1939	32	1949	144
1940	28	1950	159
1941	23	1951	133
1942	27	1952	161
1943	21	1953	226
1944	24	1954	188
1945	20	1955	189
1946	54	1956	170
1947	88	1957	154

4.4 The costs of a random sample of searches of computer databases were recorded as:

Cost (£)	6	8	10	12	14	16	18	20
Number of searches	1	3	6	10	8	8	4	2

4.5 The expenditure breakdowns of a library in two years were as follows:

Expenditure	Year 1	Year 2
Salaries	150,000	175,000
Books, etc.	110,000	140,000
Non-recurring charges	25,000	20,000

4.6 Five random samples of new scientific texts, each containing ten observations, provided the following data relating to cost:

Sample number	Mean cost	Standard deviation of cost
1	25.75	12.83
2	19.90	9.83
3	24.84	13.90
4	24.10	15.11
5	24.30	9.74

4.7 The number of junior books issued from a public library on a selection of days were:

76	16	61	37	54	58	23	19	46	38	16	45
84	58	50	30	61	30	47	27	17	42	44	25
38	18	10	29	50	10	36	70	45	58	50	29
26	50	31	44	54	24	34	52	20	58	59	38
21	38	54	38	54	16	37	29	30	38	22	6

Chapter 3

PRESENTATION OF DATA

3.0 Introduction

In some circumstances representation of data by measures of average, dispersion and correlation, as discussed in Chapter 2, is not sufficient and the complete collection of data needs to be presented. The method of presentation will depend on the nature of the data — whether quantitative or qualitative or a combination. Presentation may be in the form of a table or a diagram. In general, it can be said that a table provides more accurate information but is less easy for the reader to absorb, while a diagram provides a more easily absorbed presentation with a lesser degree of accuracy. An example of each type of diagram is given in the Appendix (pp.57−68).

3.1 Quantitative data

3.1.1 Tabular presentation

3.1.1.1 Frequency

Quantities of raw data tabulated as collected would be very difficult to absorb as seen from the small number of observations in Figure 3.1.

3.1 Number of connect hours per month of DIALOG files used by an information service − raw data

0.1	0.4	0.5	0.1	0.4	0.2
0.4	0.2	0.1	0.3	0.9	0.1
0.2	0.1	0.4	1.2	0.2	0.1
0.8	0.9	0.2	0.4	0.1	0.4

It is therefore normally presented in the form of a *frequency table* as shown in Figure 3.2.

3.2 Number of connect hours per month of DIALOG files used by an information service – frequency table

Number of connect hours	0.1	0.2	0.3	0.4	0.5	0.6	0.7	0.8	0.9	1.0	1.1	1.2
Number of files	7	5	1	6	1	0	0	1	2	0	0	1

A frequency table consists of two rows (or columns) of figures, one being the variable quantity, the other being the frequency with which the variable quantity is observed. The variable row (or column) can be recognized easily since it consists of a series of figures increasing in regular steps ('number of connect hours' in Figure 3.2). The figures in the other row (or column) show how frequently each value of the variable was observed, *e.g.* in Figure 3.2 a connect time of 0.4 hours was observed six times.

The presentation in Figure 3.2 shows the actual frequency of observation of each value of the variable.

3.1.1.2 Relative frequency (BSL, 3rd ed., 5)
Relative frequencies may similarly be presented as shown in Figure 3.3.

3.3 Relative frequency of connect hours per month of DIALOG files used by an information service

Number of connect hours	0.1	0.2	0.3	0.4	0.5	0.6	0.7	0.8	0.9	1.0	1.1	1.2
Relative no. of files (%)	29.2	20.8	4.2	25.0	4.2	0.0	0.0	4.2	8.3	0.0	0.0	4.2

Such a presentation would be chosen if it is wished to show the *proportion* rather than the actual number of observations which relate to each value of the variable.

3.1.1.3 Cumulative frequency (BSL, 3rd ed., 7–9)
In some circumstances it may be necessary to show how many observations, or what proportion of the observations, have a value of the variable less than or more than a particular figure. A *cumulative frequency table*, as shown in Figure 3.4 or Figure 3.5, is then the appropriate form of presentation.

3.4 A 'less than' cumulative frequency table of the connect hours per month of DIALOG files used by an information service

Number of connect hours	0.1	0.2	0.3	0.4	0.5	0.6	0.7	0.8	0.9	1.0	1.1	1.2	1.3
Cumulative no. of files	0	7	12	13	19	20	20	20	21	23	23	23	24

3.5 A 'more than' cumulative frequency table of the percentage of connect hours per month of DIALOG files used by an information service

Number of connect hours	0	0.1	0.2	0.3	0.4	0.5	0.6
Cumulative relative frequency (%)	100	70.9	50.1	45.9	20.9	16.7	16.7

Number of connect hours	0.7	0.8	0.9	1.0	1.1	1.2
Cumulative relative frequency (%)	16.7	12.5	4.2	4.2	4.2	0

For example, Figure 3.4 shows that 19 files had a connect time of less than 0.5 hours, whilst Figure 3.5 shows that 45.9% of the files had a connect time of more than 0.3 hours.

3.1.1.4 Grouped frequency (BSL, 3rd ed., 9–12)
If data have to be grouped for the reasons given in Section 2.1.10 (p.17), tables similar to those described in the above paragraphs can be constructed. Figure 2.9 (p.18) is, in fact, a grouped frequency table of the data given in Figure 2.8 (p.18).

3.1.2 Diagrammatic presentation
There are three types of diagram which may be of particular interest, a *histogram*, a *frequency polygon* and a *cumulative frequency graph*, all of which can be used whether the data are grouped or not.

3.1.2.1 Histogram (see Appendices (i) and (ii), and BSL, 3rd ed., 6, 11, 13)
A histogram is the most straightforward diagrammatic method of presenting quantitative data such as that in Figure 3.2 and is useful for illustrating the range and dispersion of the data.

It is easy to find the mode or modal class from the histogram. The median may also be estimated as that value of the variable for which the total areas of the columns on either side are equal.

3.1.2.2 Cumulative frequency graph (see Appendices (iii) and (iv), and BSL, 3rd ed., 9–11)

If the exact values of the median and quartiles are required, the cumulative frequency graph is a better method of presentation since those values can be read directly from it. This form of presentation is also more appropriate if 'running totals' of data are of primary interest. The data from Figures 3.4 and 3.5 can be plotted to produce graphs of this type.

3.1.2.3 Frequency polygon (see Appendix (v) and BSL, 3rd ed., 6–7)

Two histograms cannot be superimposed with clarity. Therefore, if two or more sets of quantitative data are to be compared, a *frequency polygon* should be used instead. Figure 3.6 is an example of a collection of data for which this form of presentation would be the most appropriate.

3.6 The number of words per line on two pages of a book

Number of words per line	1	2	3	4	5	6	7	8	9	10	11	12	13	14
Page 1 number of lines	1	0	0	0	0	0	3	1	9	12	6	4	4	0
Page 2 number of lines	0	0	1	0	0	1	2	4	9	8	5	6	2	1

3.2 Qualitative data

3.2.1 Tabular presentation

Quantities of raw qualitative data tabulated as collected (*e.g.* Figure 3.7) would be just as difficult to absorb as raw quantitative data.

F	N	F	F	N
F	F	N	F	F
N	F	F	N	F
F	N	F	N	N
F	F	N	F	F

3.7 Fiction (F) and non-fiction (N) books issued by a library in a given period of time – raw data

35

In this case tabulation merely involves recording the number of observations in each category as shown in Figure 3.8. The data would immediately appear in this form if the data were collected in the form of a tally chart, (see Figure 1.3, p.4).

3.8 Fiction and non-fiction books issued by a library in a given period of time

Type of book	Number of issues
Fiction	16
Non-fiction	9

If the proportions of each category are of interest, rather than actual numbers, the figures can easily be presented as percentages (Figure 3.9).

3.9 Proportions of fiction and non-fiction issued by a library in a given period of time

Type of book	Proportion of each type of issue (%)
Fiction	64
Non-fiction	36

3.2.2 Diagrammatic presentation
There are a number of types of diagram which may be particularly useful for qualitative data: a *simple column chart*, a *simple bar chart*, a *component column/bar chart*, a *grouped column/bar chart* and a *pie chart*.

3.2.2.1 Simple column chart (see Appendix (vi) and BSL, 3rd ed., 96−7)
If the data are allocated to categories of date or time as in Figure 3.10, a simple column chart is the preferred form of presentation. Each vertical column represents the number of observations falling into the relevant category.

3.10 Number of literature searches carried out by an information service over a period of years

Year	Number of searches
1984	14
1985	27
1986	32
1987	39

Since all the columns stand on the same base line, their heights can be compared easily to assess the relative quantities in the various categories.

3.2.2.2 *Simple bar chart* (BSL, 3rd ed., 99–100)

If the categories do not relate to date or time (as in Figure 3.8), a bar chart is preferred. In principle, this is the same as a column chart but consists of horizontal bars instead of vertical columns.

3.2.2.3 *Component column/bar chart* (see Appendices (vii) and (viii), and BSL, 3rd ed., 98, 101)

Collections of qualitative data can be complex and therefore require a more complex diagram than a simple bar or column chart. For example, you may need to present the fiction/non-fiction issues in two libraries. In tabular form, the data may be given as shown in Figure 3.11.

3.11 Fiction and non-fiction books issued by two libraries in a given period of time

Library	Type of book	Number of issues
Library A	Fiction	16
	Non-fiction	9
Library B	Fiction	40
	Non-fiction	20

In this case, it would be appropriate to have a single bar for each library, with each bar divided into fiction and non-fiction components. If one is interested in the proportions, rather than the actual numbers, of fiction and non-fiction issued by the two libraries, the figures can be converted to percentages and a *100% bar chart* can be drawn. Similarly, a *100% column chart* may be drawn for data in time categories (BSL, 3rd ed., 97, 99).

Thus component column and bar charts are used to show the total number of observations in each category and how that total is broken down into its component sub-categories.

3.2.2.4 *Grouped column/bar chart* (see Appendix (ix) and BSL, 3rd ed., 98–100)

It may be difficult to compare the lengths of the corresponding components in the various columns or bars of a component column or bar chart. For an easier comparison, a *grouped* column or bar chart may be preferred. Every component is shown as a separate column or bar extending from a common base line. Whilst this form of presentation simplifies the comparison of all components, it does not immediately show the total number of observations in each category.

3.2.2.5 *Pie chart* (see Appendix (x) and BSL, 3rd ed., 100–102)
A pie chart is a visually attractive form of presentation of qualitative data. It is commonly used to show the breakdown of a single collection of data into its components (*e.g.* the data in Figure 2.11 or Figure 3.8). It is not easy to compare two or more pie charts, especially if each set of data consists of more than two or three components. If two or more sets of such data are required in a presentation, some form of column or bar chart is preferred.

3.3 Qualitative/quantitative data

3.3.1 *Tabular presentation*
A table of such data consists of a column for the various categories of the qualitative variable and a column for the corresponding values of the quantitative variable. As noted in Section 1.2.3.1 (p.5), the qualitative variable is often date or time — Example 5 in that section shows such a table.

There may be more than one quantitative observation for each qualitative category, *e.g.* non-fiction, fiction, junior and total book issues for each month as illustrated in Figure 3.12.

3.12 Monthly totals of books issued from a library

Month	Non-fiction	Fiction	Junior	Total
Jan	465	3216	713	4394
Feb	513	3215	686	4414
Mar	425	3126	996	4547

3.3.2 *Diagrammatic presentation*
Line graphs and *surface graphs* are the two main forms of presentation of this type of data.

3.3.2.1 *Line graphs* (see Appendix (xi) and BSL, 3rd ed., 115–17)
In a line graph, the qualitative variable is always plotted on the horizontal axis and the quantitative variable is plotted on the vertical axis. Data for several quantitative variables (*e.g.* non-fiction, fiction and junior in

38

Figure 3.12) can be plotted on the same graph and the values for the total and each component can be read directly from the vertical scale.

3.3.2.2 *Semi-logarithmic graphs* (see Appendices (xii) and (xiii), and BSL, 3rd ed., 116—19)

On this type of graph the qualitative variable is plotted on a linear scale on the horizontal axis and the quantitative variable is plotted on a logarithmic scale on the vertical axis. It is particularly useful if the quantitative variable shows a rate of growth or decline which changes very appreciably over the time scale.

From Figure 3.12 it can be seen that the total number of issues shows a gentle increase over the months recorded and an ordinary line graph with both scales linear would be appropriate in that case. On the other hand, the data in Figure 3.13 (below) show an explosive growth. Whilst an ordinary line graph with both scales linear would illustrate the nature of that growth, values of the quantitative variable (*i.e.* the number of journals) could not be read from such a graph with equal accuracy at all parts of the time scale. If that facility is required, a semi-logarithmic graph should be used.

3.13   Growth in number of scientific journals

Date	Number of journals
1700	8
1750	10
1800	90
1850	1 000
1900	10 000
1950	85 000

3.3.2.3 *Surface graphs* (see Appendices (xiv) and (xv), and BSL, 3rd ed., 120)

When a total quantity is made up of two or more components (as in Figure 3.12) and one wishes to be able to read easily the values of the total and of each individual component then, as indicated in Section 3.3.2.1, a line graph is the appropriate choice − or a semi-logarithmic graph if the data justify it. If, however, the aim is to show how the total quantity is made up from the various components, a *surface graph* should be used. It is not easy to determine from such a graph the values of the individual components (except the component represented by the lowest 'surface'), but that is the price paid for more clearly displaying a breakdown of the total.

If one is interested in the proportions rather than the actual numbers of the various components which make up the total, the figures can be converted to percentages and a *100% surface graph* can be drawn (BSL, 3rd ed., 120–1).

A surface graph of a collection of data, such as that in Figure 3.12, may be compared with a component column chart. The surface graph would be used if the number of date categories is large; a component column chart would be preferred if the number of date categories is small.

3.4 Quantitative/quantitative data

3.4.1 Tabular presentation

3.4.1.1 *Simple table*
As noted in Section 1.2.3.2 (p.6), each observation of this type involves the recording of two quantitative variables. A table will therefore contain two columns, one for each of the two variables. There may be a third column containing numbers to identify the pairs of observations. Thus, in Figure 3.14, document number five was three years old and had been used 26 times.

3.14 Frequency of use of a number of documents of different ages

Document number	Age of document (years)	Frequency of use (times/year)
1	1	40
2	3	18
3	2	30
4	4	21
5	3	26
6	5	10
7	4	13
8	3	35

3.4.1.2 *Correlation table* (BSL, 3rd ed., 150–2, 164–5)
The simple table of Figure 3.14 has no pattern or order, other than the order in which the observations were made. Even with so few observations the data are difficult to absorb. As the number of observations grows, it becomes more and more difficult to see any underlying pattern. Any such pattern is brought out in a *correlation table*. The data from Figure 3.14 are retabulated in such a way in Figure 3.15.

3.15 Correlation table of age and frequency of use of documents

Frequency of use (times per year)	Age of document (Years)					
	1	2	3	4	5	Total
1−10					1	1
11−20			1	1		2
21−30		1	1	1		3
31−40	1		1			2
Total	1	1	3	2	1	8

From the table in Figure 3.15 it is easily seen that the older the document, the less frequently it is used.

3.4.2 Diagrammatic presentation

A *scatter diagram* is the only method of presenting this type of data. Each point on the diagram corresponds to an observation and the values of the two variables can be read from the horizontal and vertical scales of the diagram. (See Appendix (xvi) and BSL, 3rd ed., 149−50, 157−8.)

Exercises

Exercise 5

Decide which type of table and/or diagram would be most suitable for each of the following collections of data, bearing in mind constraints where stipulated.

5.1 In a survey of library users, their occupations were noted as follows:

Occupation	Number
Employed	576
Housewife	599
Pupil	215
Student	43
Unemployed	156
Retired	600

Illustrate how the library membership is made up from the occupational groups.

5.2 The number of volumes of fiction and non-fiction issued monthly to housebound readers by a County Library were recorded. The results were as follows:

Year	1982				1983							
Month	Sep	Oct	Nov	Dec	Jan	Feb	Mar	Apr	May	Jun	Jul	Aug
Non-fiction	60	101	128	157	151	162	163	114	180	104	161	207
Fiction	62	192	327	421	410	487	486	474	660	390	497	786

(i) Illustrate the relationship between the number of fiction issues and the number of non-fiction issues.

(ii) Show how the number of
(a) non-fiction issues,
(b) fiction issues,
(c) total number of issues
changed over the 12-month period.

(iii) Display the change in proportional breakdown of the monthly issues into non-fiction and fiction over the 12-month period.

5.3 The age groups of users of a mobile library service on different routes are given in the following table:

	Age group					
Route	20−29	30−39	40−49	50−59	60−69	70+
1	3	4	4	7	6	5
2	2	5	4	8	4	3
3	2	5	6	5	18	9

(i) Present the data for Route 1 in a diagram from which the modal age can easily be determined.

(ii) Draw a diagram to show more clearly any differences between the distribution of the users' ages on the three routes.

5.4 The number of connect hours to various DIALOG files on a monthly invoice were:

0.433	0.374	0.126	1.741	0.230
0.235	0.458	0.816	0.439	0.267
0.457	1.200	0.065	0.928	0.135
0.100	0.247	0.963	0.458	0.477
0.518	0.206	0.125	0.134	0.134

(i) Draw up a table to display the data more clearly.

(ii) Draw a diagram from which one may determine the median number of contact hours and also the percentage of files for which contact exceeds a given time.

5.5 Estimates of expenditure from a number of library authorities several years ago are given in the following table:

Library authority	Population	Estimates (£) Total	Estimates (£) Books	Estimates (£) Binding
A	277 100	145 932	30 400	9 600
B	158 700	63 355	8 000	1 800
C	96 100	148 065	33 000	13 000
D	69 700	20 165	4 750	1 100
E	52 300	42 030	11 000	1 700
F	21 820	3 702	700	200

(i) Draw a diagram to show the estimated total amounts spent per head of population by the various authorities and how these amounts are broken down into expenditure on books, binding and other items, respectively.

(ii) If the primary interest was to determine the authority which allocated the highest proportion of its estimate to books, what form of diagram would be preferable?

5.6 Given the following figures for the UK publishing output in three areas of the humanities, draw a diagram to compare the output in the three areas in each year and to compare those outputs across the years.

Subject	Year 1974	Year 1975	Year 1976
Religion	1163	1255	1168
Arts	1942	2187	2242
History	1975	2086	1992

Chapter 4

MANIPULATION OF DATA

4.0 Introduction

Chapters 2 and 3 are concerned with methods of communicating information about data. This chapter is concerned with methods of using statistical data in order to draw conclusions which may be of value for research or managerial purposes. For example, information about populations may be deduced from samples drawn from them (this was introduced in Section 2.1.11); tests may be performed to determine the significance of any differences between sets of data; or calculations may be performed to derive further information based on the data collected. As for representation and presentation, the choice of method of manipulation will depend on the nature of the data, *i.e.* whether it is quantitative or qualitative or a combination.

4.1 Quantitative data

There are various procedures which can be applied to quantitative data, depending on the information available and the information being sought. They include calculations based on probability theory and various statistical tests.

4.1.1 Probability

Probability is concerned with determining, from observations made, the likelihood of a particular event happening or a particular characteristic being observed. From observations made on samples, information may be deduced concerning the likelihood of particular events or characteristics occurring in the population from which the samples were taken. Deductions may be based on various concepts, including the *standard error of the probability of success*, the *binomial distribution* and the *Poisson distribution*.

4.1.1.1 Standard error of the probability of success (BSL, 3rd ed.,
　　　　55–8)

Samples can be taken from a population. If, in taking a sample, the
number of occurrences of a characteristic of interest are counted, then
the probability of that characteristic occurring (the probability of
'success') can be determined for that sample. The probability of success
will vary quantitatively from one sample to another. Consequently there
is a standard error of the probability of success which can be estimated
from the information obtained in a sample taken. That can then be used
to obtain information on the minimum and maximum number of times
the characteristic of interest is likely to occur in the population.

This procedure is only useful when dealing with a *large population*
from which a *large sample* can be taken.

Example 13

In a given year, 30 000 books were borrowed. Out of a random sample
of 400 of those books, 40 were overdue. What is the maximum number
of overdue books which may be expected in the year? Hence what is
the maximum sum which may be received in fines in that year if the
mean fine is 10p?

The characteristic of interest is a returned book being overdue. The
information given is sufficient to find that the probability of a returned
book in the given sample being overdue (a 'success'!) is 0.1 (40/400).
The standard error of the probability of success can then be determined as

$$0.015\left(\sqrt{\frac{0.1 \times 0.9}{400}}\right).$$

Hence it can be deduced that the maximum probability of success in any
sample is 0.13 (0.1 + 2 × 0.015). Finally, it can be calculated that,
out of the population of 30 000 books returned annually, the maximum
number that are likely to be overdue is 3 900 (0.13 × 30 000). Hence,
at 10p per overdue, the income from fines is not likely to exceed £390.

4.1.1.2 Binomial distribution (BSL, 3rd ed., 59–66)

If a set of 'trials' is carried out consisting of a *small number of
observations* taken from a population, some of the observations may be
'successes' and some 'failures'. Success or failure are the only *two
possible outcomes* from such observations. The probability of success,
as noted in Section 4.1.1, can be found from observations made on a
large sample taken from the population. By using that probability of

success and the theory of the binomial distribution, the probability of there being no success in the trial − or one success, or two successes, etc. − can be determined.

Example 14

If 30 000 books are borrowed in a year and out of a random sample of 400 of those books, 40 are overdue:

what is the probability that a book returned by a borrower will be overdue?

what is the probability that a reader returning three books will have none overdue − or one overdue, or two, or three?

if there are 10 000 borrowers in a year and each returns three books, how much money could be expected in fines in that year if the mean fine is 10p?

As in Example 13, the probability of success (a book being overdue) is 0.1. The trials involve observation of the three books returned by a reader to determine whether each one is or is not overdue. By using the binomial distribution, it can be found that:

the probability that the three trials include no overdues ('successes') is 0.729;

the probability that the result of one of the three trials is a 'success', *i.e.* one out of the three returns is overdue, is 0.243;

the probability that two are overdue is 0.027;

and the probability that all three are overdue is 0.001.

As there are 10 000 readers in the year, it can be expected that there will be 7 290 readers with none overdue, 2 430 with one overdue, 270 with two overdue and 10 with three overdue. Charging each overdue at 10p would result in a total income of £300.

If the probability of success and number of observations are known, the mean and standard deviation of the binomial distribution can be determined.

Example 15

The probability that a book returned to the library will be overdue is $\frac{1}{10}$. If 100 books are returned in a day, what are the minimum and maximum number of overdues which may be expected in that time?

From the information given, the mean can be found to be 10 overdue books per day with a standard deviation of 3. Hence it can be deduced (see Section 2.1.3.2) that not less than 4 and no more than 16 of the books returned in a day are likely to be overdue.

46

4.1.1.3 *Poisson distribution* (BSL, 3rd ed., 66−7)

A special situation arises if the *number of observations* in the trials is *very large* and the *probability of success is very small*. Under these circumstances the binomial distribution tends towards the Poisson distribution. Bearing these constraints in mind, the Poisson distribution may be used in a similar way to the binomial distribution.

Example 16

Out of a random sample of 400 books returned to a library, only 20 are overdue. If a reader is allowed to borrow a large number of books, what is the probability that, when returning 40 books, there will be none overdue − or one, or two, or three?

In this case, the probability of success is very low, *viz*: 0.05 (20/400). Each trial involves 40 observations, hence the mean number of overdues out of 40 returned is found to be 2. Using the theory of the Poisson distribution, it can be deduced that:

the probability that none of the 40 will be overdue is 0.1353;

the probability of one overdue is 0.2706 as also is the probability of there being two overdue;

the probability of there being three overdue has dropped to 0.1804.

From these probabilities, it can be deduced that if 100 readers each bring back 40 books, approximately 14 will have no overdues; 27 will have one overdue; another 27 will have two overdue; 18 will have three overdue.

From the value of the mean, the standard deviation for these samples can be calculated to be $\sqrt{2}$, approximately 1.414. Hence it can be deduced that the maximum number of overdues out of 40 returned books is not likely to exceed $2 + (2 \times 1.414)$, (the mean plus two times the standard deviation; see Section 2.1.3.2) i.e. approximately 5.

4.1.2 Statistical tests

It is often necessary to compare two sets of data to determine whether they are similar or whether a significant difference exists. As far as quantitative data are concerned, *t* and *Z* tests and F tests are of interest.

4.1.2.1 *t- and Z-tests* (BSL, 3rd ed., 79−86)

t and *Z* tests are very similar and the choice simply depends on the number of observations made:

if the number of observations is *less than 30*, the test is referred to as a *t-test*;

47

if the number of observations is *30 or more*, the test is referred to as a *Z-test*.

This type of test is used if the mean of a sample and the mean of the population from which the sample has been taken are both known and it is necessary to decide whether or not those two means are significantly different.

Similarly, the significance of any difference between the means of two samples taken from a population can be tested.

Apart from the two means, the standard deviations of the samples are needed in order to perform the necessary calculations. That information can easily be obtained from the raw data, as discussed in Sections 2.1.2 and 2.1.3.2 (pp.12 and 14).

Example 17
The mean number of books received by inter-library loan by member libraries of a regional scheme (including county libraries) was 1757. The number of books received by the county libraries in the scheme were:

$$1975 \quad 2139 \quad 2654 \quad 2442 \quad 3402 \quad 4451 \quad 2521 \quad 4096$$

Is the mean number of books received by the county libraries significantly higher than the mean for all members of the scheme?

The county libraries constitute a sample from the population of all member libraries for which the mean receipts is given as 1757. From the raw data for the county libraries, the sample mean and standard deviation can be calculated as 2960 and 918.7, respectively.

The value of t can be calculated and used to show that the mean number of books received by the county libraries is significantly greater than the mean number of books received by all members of the scheme.

Example 18
The number of issues of junior non-fiction on a random sample of days in May and November were as shown in Figure 4.1.

4.1 Junior non-fiction issues

May		Nov	
36	37	34	78
28	97	89	89
32	37	22	34
39	33	44	22
27	114	49	33
114	35	33	17

Does there appear to be a significant difference in demand between the two months?

Here, there are two samples and it will be necessary to decide if there is any significant difference between the means of the two samples.

The means can be calculated from the raw data to be 52.42 for May and 45.33 for November. Also from the raw data, the standard deviations can be found to be 34.16 and 25.85, respectively. These figures can be used in a t-test to deduce that there appears to be no significant difference between the average demand for junior non-fiction in May and the average demand in November.

4.1.2.2 F-test (BSL, 3rd ed., 87−8)

As discussed in Section 2.1.3 (p.13), dispersion is an important measure of quantitative data. In comparing two samples of data, it may be important to know whether the dispersions are comparable or significantly different. The *variance*, which is *the square of the standard deviation*, is used as a measure of the dispersion and utilized in an F-test to decide whether or not the two samples may be regarded as having been drawn from the same population.

In Example 18, the standard deviations were found to be 34.16 and 25.85, respectively, giving the corresponding variances as 1166.9 and 668.2. Using these values in an F-test, it is found that the F value is not significant. Hence it may be deduced that the variances are not significantly different. The extent of the variation in book issues in November is not significantly different from the extent of the variation in May.

4.2 Qualitative data

Under this heading there is only one manipulation to consider, *viz*: the *chi-squared test*. (The greek letter chi, χ, is often used when referring to this test, *i.e.* it may be written as the 'χ-squared test' or 'χ^2 test'.)

4.2.1 Chi-squared test (BSL, 3rd ed., 105−10)

The chi-squared test is used to decide whether observed frequencies of occurrence of a qualitative attribute differ significantly from the expected or theoretical frequencies.

Example 19

Example 6 (p.6) gave data on the number of daily issues of junior non-fiction from a library during the course of a week. It might be hypothesized that the number of issues may be expected to be independent of the day of the week. In that case, one sixth of the week's total may be expected to be issued each day as shown in Figure 4.2.

4.2 Actual and expected junior non-fiction issues

Day	Actual issues	Expected issues
Mon	39	42.17
Tue	14	42.17
Wed	21	42.17
Thu	47	42.17
Fri	36	42.17
Sat	96	42.17
Total	253	253(.02)

A chi-squared test is used to show whether there is any significant difference between the actual daily issues and the expected issues. In fact, it will show that there is a highly significant difference between the actual and expected issues, indicating that the hypothesis is incorrect and that the number of issues is very dependent on the day of the week.

4.2.1.1 Contingency tables
Some situations are more complex and require the production of a *contingency table*. The principle remains the same — comparison is made between a contingency table of observed frequencies and a corresponding contingency table of expected frequencies.

Example 20
In 1984−5, a library authority spent £550 000 on staff, £230 000 on books and £140 000 on other items. In 1987−8, the authority spent £810 000 on staff, £330 000 on books and £210 000 on other items. Did the pattern of expenditure change significantly between 1984−5 and 1987−8?

The observed data can be compiled into a contingency table as shown in Figure 4.3.

4.3 Contingency table of observed frequencies

Year		Expenditure (£'000s)		
	Staff	Books	Other	Total
1984−5	550	230	140	920
1987−8	810	330	210	1350
Total	1360	560	350	2270

A table of expected frequencies can be deduced as shown in Figure 4.4.

4.4 Contingency table of expected frequencies

Year		Expenditure (£'000s)		
	Staff	Books	Other	Total
1984−5	551.19	226.96	141.85	920
1987−8	508.81	333.04	208.15	1350
Total	1360	560	350	2270

In this case, a chi-squared test shows that there appears to be no significant difference between the observed and expected frequencies and hence it can be concluded that there is no appreciable change in the pattern of expenditure.

4.3 Qualitative/quantitative data

A comparison of the degree of dispersion of quantitative data in several qualitative categories involves a procedure known as *analysis of variance.*

4.3.1 Analysis of variance (BSL, 3rd ed., 142−4)

The purpose of analysis of variance is perhaps best appreciated from a specific example:

Example 21

The number of books stored per shelf in a library may be of interest. If a random sample of shelves is selected and the number of books on each shelf are counted, the quantitative data collected can be presented in a frequency table as shown in Figure 4.5.

4.5 **Frequency table showing variation of number of books stored per shelf in a random sample of shelves**

Books per shelf	Number of shelves	Books per shelf	Number of shelves
16	1	30	4
17	0	31	0
18	0	32	1
19	0	33	3
20	0	34	1
21	3	35	0
22	0	36	2
23	4	37	0
24	0	38	0
25	4	39	0
26	3	40	0
27	0	41	0
28	2	42	0
29	1	43	1

There is obviously quite a variation in the number of books stored per shelf (this will depend on the sizes of the individual books). Another factor which may affect the degree of variation is the subject field. If the data from Figure 4.5 are separated according to subject categories, the frequency table in Figure 4.6 is obtained.

It is clear from Figure 4.6 that:

(i)   there is variation within each subject field;

(ii)   the extent of the variation differs from one subject field to another;

(iii)   the extent of the variation in each individual subject field is less than the extent of the variation in the overall sample (the 'totals' column;

(iv)   the location of the data varies from one subject field to another (*e.g.* Law lies between 16 and 33, while Production lies between 23 and 43).

The total variability of shelf storage capacity may therefore legitimately be divided into two parts, one part associated with differences between books in the same subject field and the other part associated with differences between subject fields. Analysis of the variability, including a final F-test (see Section 4.1.2.2), enables a conclusion to be reached as to whether or not the storage capacity of shelves depends significantly on the subject field.

52

4.6 Frequency table showing variation of number of books stored per shelf according to subject category

Books per shelf	Number of shelves			
	Geography	Law	Production	Total
16		1		1
17				0
18				0
19				0
20				0
21		3		3
22				0
23	1	1	2	4
24				0
25	4			4
26		3		3
27				0
28	1	1		2
29	1			1
30	2		2	4
31				0
32			1	1
33	1	1	1	3
34			1	1
35				0
36			2	2
37				0
38				0
39				0
40				0
41				0
42				0
43			1	1

4.4 Quantitative/quantitative data

The nature of data in this category, involving two related variables, has been indicated in Sections 1.4, 2.4 and 3.4. The procedure for manipulating the observed data to derive further, unobserved, information is known as *regression*.

4.4.1 Regression (BSL, 3rd ed., 167–72)

If the value of one of the observed variables is specified, the problem is to determine the expected corresponding value of the other variable. For example, from the data of Figure 3.14 (p.40), you may be required to estimate the frequency of use of a document which is (say) six years old; or, using the data of Example 11 (p.28), you may wish to estimate the cost of a volume containing (say) 6000 abstracts. Conversely, an estimate of the number of abstracts which may be expected in a volume costing (say) £90 may be sought. All such estimates may be obtained by means of regression calculations.

Exercises

Exercise 6
Decide which type of manipulation of the data would be possible in each of the following examples.

6.1 On average, only one carrel out of three available in a library is free at any time. What is the probability that a reader will have to wait for a free carrel?

6.2 Abstracts in *Economics Abstracts* are written in English, French and German. The number of words in random samples of eight abstracts in each language were recorded as follows:

English	French	German
71	111	67
118	113	75
52	84	61
47	84	99
59	84	58
65	94	107
84	90	113
111	90	95

Does the length of abstract appear to depend on the language in which it is written?

6.3 The cost of the most popular program cassettes for a micro-computer were (£.p):

7.95 7.95 5.95 5.95 5.50 6.95 5.95 4.95 6.50 4.95

Is the mean cost of this sample significantly different from the mean cost of all the cassettes available for the machine, which was £6.68?

6.4 In a special library containing 5000 volumes, 500 of the books were selected at random and, of those, only 300 had been used in the course of the past 12 months. Assuming that you can store 25 books per shelf, what is the minimum amount of shelves you may expect to empty if you disposed of all the unused books in the library?

6.5 The number of volumes of fiction and non-fiction issued monthly to housebound readers by a county library were:

Year	1982												1983
Month	Sep	Oct	Nov	Dec	Jan	Feb	Mar	Apr	May	Jun	Jul	Aug	Sep
Non-fiction	60	101	128	157	151	162	163	114	180	104	161	207	204
Fiction	62	192	327	421	410	487	486	474	660	390	497	786	788

(i) Is there any significant difference in the variability of fiction and non-fiction issues?

(ii) In a month when there are 140 non-fiction issues, how many fiction issues would you expect?

6.6 Searching a post-coordinate file is found to result in 3% of the items retrieved being 'false drops'. If a specific search for information produces 60 references, what is the probability that there will be no false drops?

6.7 UK output figures for three areas of the humanities between 1974 and 1976 were as follows:

Subject	Year		
	1974	1975	1976
Religion	1163	1255	1168
Arts	1942	2187	2242
History	1975	2086	1992

Did the subject pattern of the output change significantly between 1974 and 1976?

6.8 In an investigation of the efficiency of the information retrieval process, the following data were obtained:

Search number	Total documents retrieved	Relevant documents retrieved
1	79	21
2	18	10
3	20	11
4	123	48
5	16	8
6	109	48
7	48	25
8	2	1
9	36	5

In a search in which 60 documents are retrieved, how many would be expected to be relevant?

6.9 If 10% of library users seek the librarian's help, what is the probability that two or more of the 30 readers entering the library in a day are going to seek help?

6.10 The following data are the prices of volumes in two subject fields, A and B, advertised in a publisher's list. Is there any significant difference between the average prices of books in the two subjects?

A			B		
19.95	45.00	27.50	25.00	55.00	15.00
9.95	37.50	9.50	14.50	24.00	9.50
16.00	5.95	35.00	22.50	40.00	17.00
17.50	15.00	32.50	17.50	22.50	21.00
17.00	15.00	27.50	35.00	55.00	18.00
17.00	20.50	9.95	32.50	20.50	14.50
21.50	55.00	25.00	15.00	45.00	30.00
7.50	12.00	30.00	20.00	24.00	30.00
30.00	16.00	22.50	45.00	15.00	16.00
18.50	7.50	21.00	18.50	20.00	32.50

6.11 A typist makes, on average, two mistakes per page. On how many pages of a 50 page report would you expect to find errors?

6.12 A bibliography contains 2055 references. Out of a random sample of 95 of those references, 11 were found to be in a foreign language. What is the maximum number of foreign language items that you might expect to find in the whole bibliography?

Appendix

EXAMPLES OF DIAGRAMMATIC PRESENTATION OF DATA

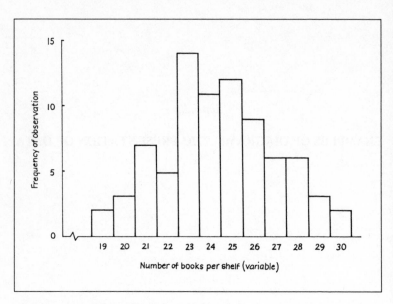

(i) Number of books per shelf – histogram

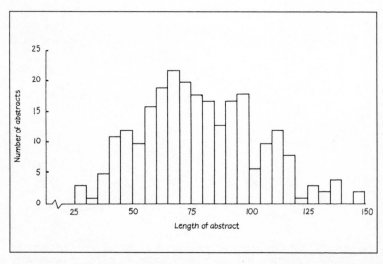

(ii) Lengths of abstracts in *Economics Abstracts*, July 1972 – histogram

58

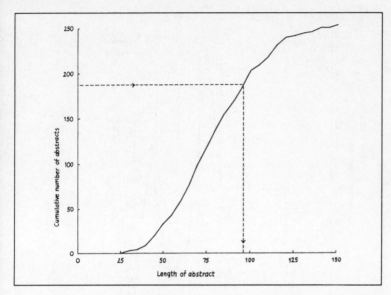

(iii) Length of abstracts in *Economics Abstracts*, July 1972 – cumulative frequency graph ('less than')

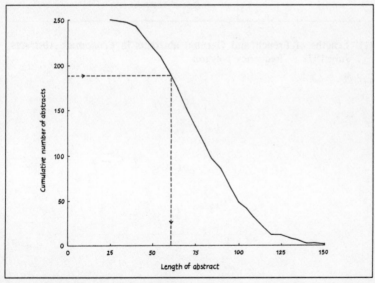

(iv) Length of abstracts in *Economics Abstracts*, July 1972 – cumulative frequency graph ('more than')

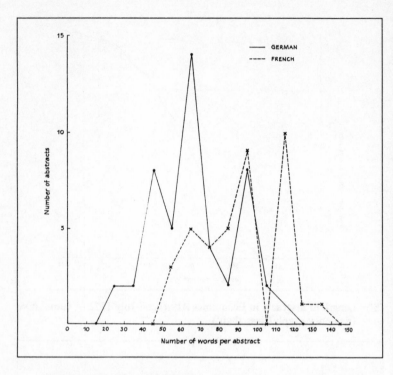

(v) Lengths of French and German abstracts in *Economics Abstracts*, July 1972 − frequency polygon

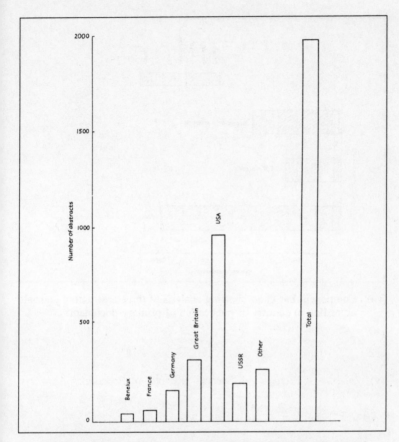

(vi) Simple column chart showing analysis of *Computer and control abstracts*, July 1971, according to country of publication of primary document

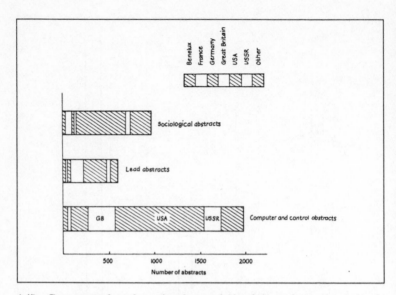

(vii) Component bar chart showing analysis of three abstracting journals
       according to country of publication of primary document

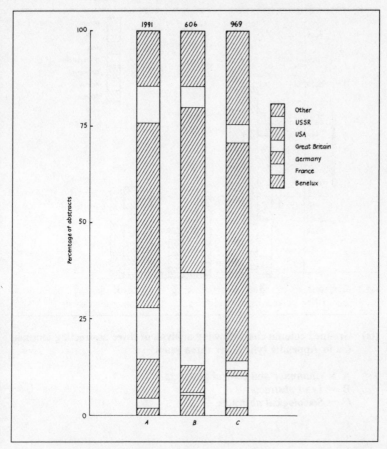

(viii) 100% component column chart showing analysis of three abstracting
        journals according to country of publication of primary document
        in 1971

A = *Computer and control abstracts*
B = *Lead abstracts*
.C = *Sociological abstracts*

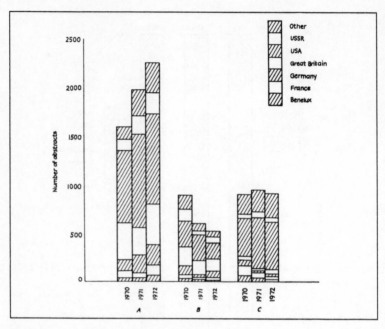

(ix) Grouped column chart showing analysis of three abstracting journals
(as in Appendix (viii)) over three years

A = *Computer and control abstracts*
B = *Lead abstracts*
C = *Sociological abstracts*

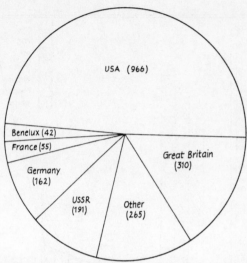

(x) Pie chart of abstracts in *Computer and control abstracts*, July 1971, according to country of publication of primary document

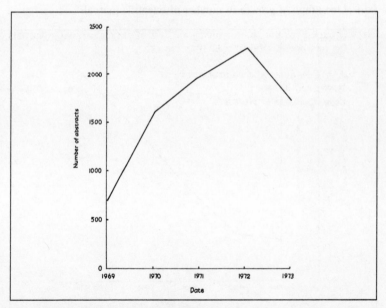

(xi) Line graph showing the number of abstracts in the July issues of *Computer and control abstracts*

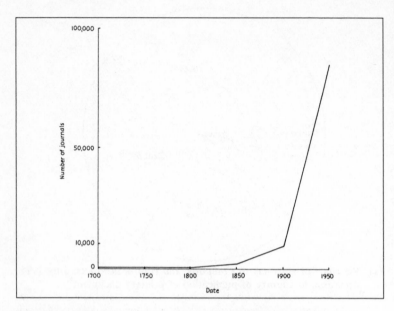

(xii)   Line graph of growth in number of scientific journals

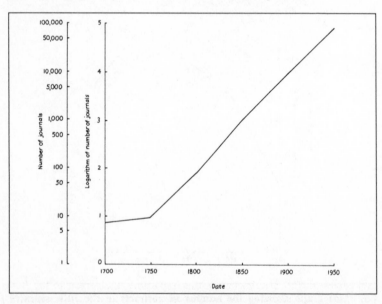

(xiii)   Semi-logarithmic graph of the growth of number of scientific journals

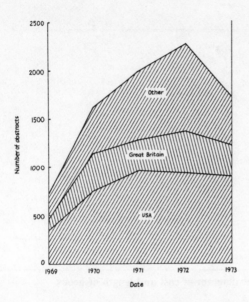

(xiv) Surface graph showing the number of abstracts in the July issues of *Computer and control abstracts* broken down by country of origin

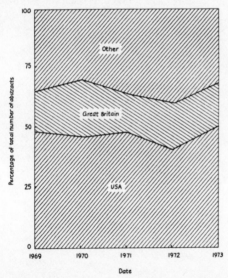

(xv) 100% surface graph showing the proportions of abstracts in the July issues of *Computer and control abstracts* broken down by country of origin

67

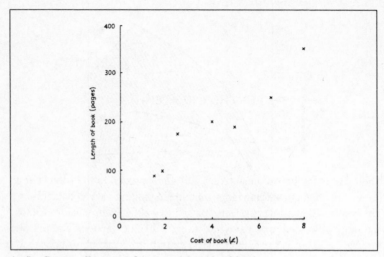

(xvi)   Scatter diagram of cost and length of books

FURTHER READING

There are many books on statistics and the approach to the subject and level of mathematical knowledge and understanding vary considerably. The books in this list were selected as being of a more introductory nature and more advanced texts have intentionally been excluded. The reason for such an extensive list is that individual libraries and bookshops tend to have a limited collection of titles but it is hoped that at least some of those listed will be within reach of most readers. Also it will give a reader the opportunity to browse to find an approach to the subject which he or she finds most comprehensible. A treatment that suits one person may not suit another. Many of the books listed indicate that a minimal amount of mathematical knowledge is required but items 7 and 41 are the only two written with the needs of librarians and information scientists specifically in mind.

1 Anderson, A. J. B., *Interpreting data: a first course in statistics*, London, Chapman and Hall, 1989.
2 Bartz, A. E., *Basic statistical concepts*, 3rd ed., New York, Macmillan, 1988.
3 Bevan, J. M., *Introduction to statistics*, London, Newnes, 1968.
4 Booth, D. J., *A first course in statistics*, Eastleigh, DP Publications, 1987.
5 Bryars, D. A., *Advanced level statistics*, London, Bell and Hyman, 1983.
6 Cain, R. B., *Elementary statistical concepts*, Philadelphia, W. B. Saunders, 1972.
7 Carpenter, R. L. and Vasu, E. S., *Statistical methods for librarians*, Chicago, American Library Association, 1978.
8 Caswell, F., *Success in statistics*, London, John Murray, 1982.

9   Caulcott, E., *Significance tests*, London, Routledge and Kegan Paul, 1973.

10 Cohen, S. S., *Practical statistics*, London, Arnold, 1988.

11 Conner, L. R. and Morrell, A. J. H., *Statistics in theory and practice*, 7th ed., London, Pitman, 1977.

12 Downie, N. M. and Heath, R. W., *Basic statistical methods*, 5th ed., New York, Harper and Row, 1983.

13 Ehrenberg, A. S. C., *A primer in data reduction: an introductory statistics textbook*, Chichester, Wiley, 1982.

14 Erricker, B. C., *Elementary statistics*, 3rd ed., London, Hodder and Stoughton, 1981.

15 Fitz-gibbon, C. T. and Morris, L. L., *How to calculate statistics*, Beverly Hills, Sage Publications, 1978.

16 Greer, A., *A first course in statistics*, Cheltenham, Thornes, 1980.

17 Hannagan, T. J., *Mastering statistics*, 2nd ed., London, Macmillan Education, 1986.

18 Hannagan, T. J., *Workout statistics GCSE*, London, Macmillan Education, 1986.

19 Harper, W. M., *Statistics*, 5th ed., Plymouth, Macdonald and Evans, 1987.

20 Hays, S., *An outline of statistics*, 8th ed., London, Longman, 1970.

21 Hayslett, H. T., *Statistics made simple*, 3rd ed., Oxford, Heinemann, 1974.

22 Hine, J. and Wetherill, G. B., *A programmed text in statistics*, London, Chapman and Hall, 1975.

23 Hoel, P. G., *Elementary statistics*, 4th ed., New York, Wiley, 1976.

24 Ilersic, A. R., *Statistics*, 14th ed., London, HFL (Publishers), 1980.

25 Koosis, D., *Statistics: a self-teaching guide*, 3rd ed., New York, Wiley, 1985.

26 Kuebler, R. R. and Smith, H., *Statistics: a beginning*, New York, Wiley, 1976.

27 Loveday, R., *Practical statistics and probability*, Cambridge, Cambridge University Press, 1974.

28 Lucas, H., *Statistical methods*, London, Butterworth, 1970.

29 Moore, D. S., *Statistics concepts and controversies*, 2nd ed., New York, W. H. Freeman, 1985.

30 Newson, J. and Matthews, M., *The language of basic statistics*, London, Longman, 1971.

31 Owen, F. and Jones, R. H., *Statistics*, 2nd ed., Stockport, Polytech Publishers, 1982.

32 Phillips, J. L., *How to think about statistics*, 3rd ed., New York, W. H. Freeman, 1988.

33 Plews, A. M., *Introductory statistics*, London, Heinemann Educational, 1979.

34 Porkess, R., *Dictionary of statistics*, London, Collins, 1988.

35 Rees, D. G., *Essential statistics*, London, Chapman and Hall, 1985.

36 Reichmann, W. J., *Use and abuse of statistics*, London, Chapman and Hall, 1962.

37 Rowntree, D., *Statistics without tears*, Harmondsworth, Penguin, 1981.

38 Runyon, R. P., *Winning with statistics*, Reading, Mass., Addison-Wesley, 1977.

39 Sanders, D. H., Eng, R. J. and Murph, A. F., *Statistics: a fresh approach*, 3rd ed., New York, McGraw Hill, 1985.

40 Schutte, J. G., *Everything you always wanted to know about elementary statistics (but were afraid to ask)*, Englewood Cliffs, Prentice-Hall, 1977.

41 Simpson, I. S., *Basic statistics for librarians*, 3rd ed., London, Library Association Publishing, 1988.

42 Swinscow, T. D. V., *Statistics at square one*, 8th ed., London, British Medical Association, 1983.

43 Walker, J. A. and McLean, M. M., *Ordinary statistics*, 2nd ed., London, Arnold, 1983.

44 Wright, H., *Statistics for GCSE*, London, Pitman, 1988.

ANSWERS TO EXERCISES

Exercise 1
1.1 Quantitative: the number of centimetres in the height of each book observed can be counted.

1.2 Qualitative: each book is allocated to a subject category, *e.g.* History, Physics, Geography.

1.3 Qualitative: each copy is allocated to a category such as catalogue reproduction, inter-library loan, copying for reader.

1.4 Quantitative: the number of pounds required to buy each book can be counted.

1.5 Qualitative: each pound spent is allocated to a category, *e.g.* staff, stock, administration.

1.6 Quantitative: for each library observed, the number of readers served can be counted.

1.7 Qualitative: each reader observed can be allocated to the area in which he or she lives.

1.8 Quantitative: the number of references in each bibliography observed can be counted.

1.9 Qualitative: each library user can be allocated to a category according to whether he telephones, writes a letter or makes a personal visit.

1.10 Qualitative: authors are either male or female.

Exercise 2
2.1 Age is number of years since publication (quantitative) and frequency of use is number of times used in a specific period of time (quantitative)

2.2 Both the number of volumes in stock and the number issued per day can be counted (quantitative/quantitative)

2.3 The date is qualitative; the number of issues in the month can be counted (quantitative).

2.4 The year is qualitative. Staff and stock are qualitative attributes of an item of expenditure; the number of pounds spent in a year on staff or stock can be counted (quantitative).

2.5 Both the number of items obtained and number of items supplied by the library can be counted (quantitative/quantitative).

2.6 The number of periodicals (quantitative) published each year (qualitative).

2.7 Both the number of fiction issues and the number of non-fiction issues in the specified time can be counted (quantitative/quantitative).

2.8 The number of fiction issues can be counted (quantitative) for each year (qualitative).

2.9 For each volume observed, the number of pounds required to pay for it can be counted and the number of abstracts contained in it can be counted (quantitative/quantitative).

2.10 The average number of minutes taken to carry out a search can be counted (quantitative) after regular intervals of time − *e.g.* daily (quantitative).

Exercise 3

3.1 For each library observed, the population was counted and the stock was counted (quantitative/quantitative).

3.2 Each library user was allocated to the most appropriate occupational category. Nothing was counted on observation. The number in each category was counted on completion of the data collection (qualitative).

3.3 The categories of date are immediately obvious. For each year observed, the number of reviews published and the number of patents published were recorded. These are both quantitative variables although not related. This is therefore a two-fold collection of qualitative/quantitative data.

3.4 For each stopword observed, the number of characters it contained were counted (quantitative).

Exercise 4

4.1 Age and frequency are both quantitative. The Pearson correlation coefficient will indicate the degree of relationship between the two variables.

4.2 Language is qualitative. There is no more concise form than that given.

4.3 These are qualitative/quantitative data. A moving average will smooth out erratic variations to show the general trend.

4.4 Cost of a search is quantitative. The mode would indicate the cost which is experienced most frequently. The mean would enable an estimate for next year's budget to be obtained. The range and/or standard deviation would indicate the variability in cost of searches. The median and quartiles would divide the searches into four equal groups regarding cost.

4.5 The type of expense is qualitative; the amount spent in a given year is quantitative. Simple or weighted indexes will compare the costs in year two with those in year one.

4.6 Cost is a quantitative variable. An estimate of the maximum value of the mean cost and the standard deviation for all texts will help in preparing a budget.

4.7 The data are quantitative. With such a range of values, grouping of the data is necessary. The modal class will give an indication of the most common number of daily issues.

Exercise 5

5.1 The data are qualitative since each user observed was allocated to one of the six categories. There is only one set of data for which the total is to be shown made up from its components. Hence a pie chart would be a good choice. An alternative would be a simple bar chart, bars being preferred since the categories do not relate to time.

5.2 For each month (qualitative), the number of non-fiction and the number of fiction issues were counted. There are therefore two quantitative values − possibly related − for each month observed.

    (i) The relationship, if any, between the number of non-fiction issues and number of fiction issues can be illustrated by means of either a correlation table or a scatter diagram.

    (ii) The relationship between each quantitative variable and the month can be displayed in a line graph. There is no explosive growth or decline in number of issues so a semi-logarithmic presentation would be inappropriate.

    (iii) To display a breakdown of the total issues, a surface graph is required and since a *proportional* breakdown is specified, a 100% surface graph is necessary. A component column chart might be considered, but since so many months are involved, it would not be as suitable as the surface graph.

5.3　The number of years in the age of each user can be counted and is therefore a quantitative variable. There are three sets of data presented in the form of a frequency table.

(i)　Since only one set of data is to be presented, a histogram can be used. The tallest column will give the modal value.

(ii)　A frequency polygon may be used as an alternative form of display for part (i) but it is essential to use such a diagram for part (ii) where three sets of quantitative data are to be compared.

5.4　The number of connect hours can be counted for each file and is therefore quantitative.

(i)　A frequency table is the appropriate table type.

(ii)　A 'more-than' cumulative frequency graph is necessary.

5.5　For each authority, each pound in the estimates has been allocated to the category of 'books', 'binding' or 'other'. Therefore the data are qualitative.

(i)　The figures will need to be converted to £ per head. To display the total amount for each authority and the breakdown of the total into its component categories, a component bar chart is required − bars because the authority category is not related to time.

(ii)　A 100% bar chart will be necessary to show which authority allocated the highest proportion to books.

5.6　In each year observed, each publication is allocated to the appropriate subject area. Date and subject are both qualitative variables. Since total annual output does not appear to be of interest, a component bar/column chart is not necessary. Comparisons between the three areas are therefore most easily achieved by using a grouped bar/column chart. The prime interest appears to be a comparison of subject areas in each year. Each group therefore relates to a year, so a grouped column chart would be preferred to a grouped bar chart.

Exercise 6

6.1　The number of carrels can be counted (quantitative). Small number of observations (3), high probability of success (⅓): use the binomial distribution.

6.2 Number of words can be counted (quantitative); languages cannot be counted (qualitative). Qualitative/quantitative data: use analysis of variance.

6.3 Cost of cassette is quantitative. Mean of sample is to be compared with mean of population for small sample: use the t-test.

6.4 Number of books can be counted (quantitative). Probability of a book being used can be calculated from data given and hence the problem solved using the standard error of probability of success.

6.5 Numbers of non-fiction and fiction issues can be counted (quantitative).

    (i) Variance for each category can be calculated and compared by the F-test.

    (ii) Assuming the two quantitative variables are related, the problem can be solved by regression.

6.6 Number of items retrieved can be counted (quantitative). Large number of observations (60) with a small probability of success (3%): use the Poisson distribution.

6.7 Date and subject are both qualitative: use the chi-squared test.

6.8 'Total documents retrieved' and 'relevant documents retrieved' can both be counted (quantitative). Problem can be solved by regression.

6.9 Number of readers counted (quantitative). Probability of success is high enough and number of observations small enough to use the binomial distribution.

6.10 Cost can be counted (quantitative). Averages of two samples to be compared for samples of 30 observations: use the Z-test.

6.11 Mistakes can be counted (quantitative). Large number of characters to be observed and very low probability of success (*i.e.* finding a mistake): use the Poisson distribution.

6.12 Number of references can be counted (quantitative). Probability of reference being in a foreign language can be calculated from the data given and hence the problem solved via standard error of probability of success.

INDEX